The Complete Guide to *Poultry Breeds*

Everything You Need to Know Explained Simply

By Melissa Nelson

THE COMPLETE GUIDE TO POULTRY BREEDS: EVERYTHING YOU NEED TO KNOW EXPLAINED SIMPLY

1405 SW 6th Avenue • Ocala, Florida 34471 • Phone 800-814-1132 • Fax 352-622-1875
Web site: www.atlantic-pub.com • E-mail: sales@atlantic-pub.com
SAN Number: 268-1250

Library of Congress Cataloging-in-Publication Data

Nelson, Melissa G. (Melissa Gwyn), 1969-
The complete guide to poultry breeds : everything you need to know explained simply / by Melissa Nelson.
p. cm.
Includes bibliographical references and index.
ISBN-13: 978-1-60138-377-8 (alk. paper)
ISBN-10: 1-60138-377-0 (alk. paper)
1. Poultry. I. Title.
SF487.N39 2011
636.5--dc22

2010053019

Printed in the United States

PROJECT MANAGER: Shannon McCarthy
PROOFING: Hayley Love • hloveunlimited@gmail.com
INTERIOR LAYOUT: Antoinette D'Amore • addesign@videotron.ca
COVER DESIGN: Meg Buchner • meg@megbuchner.com
BACK COVER DESIGN: Jackie Miller • millerjackiej@gmail.com

We recently lost our beloved pet "Bear," who was not only our best and dearest friend but also the "Vice President of Sunshine" here at Atlantic Publishing. He did not receive a salary but worked tirelessly 24 hours a day to please his parents. Bear was a rescue dog that turned around and showered myself, my wife, Sherri, his grandparents Jean, Bob, and Nancy, and every person and animal he met (maybe not rabbits) with friendship and love. He made a lot of people smile every day.

We wanted you to know that a portion of the profits of this book will be donated to The Humane Society of the United States. ***–Douglas & Sherri Brown***

The human-animal bond is as old as human history. We cherish our animal companions for their unconditional affection and acceptance. We feel a thrill when we glimpse wild creatures in their natural habitat or in our own backyard.

Unfortunately, the human-animal bond has at times been weakened. Humans have exploited some animal species to the point of extinction.

The Humane Society of the United States makes a difference in the lives of animals here at home and worldwide. The HSUS is dedicated to creating a world where our relationship with animals is guided by compassion. We seek a truly humane society in which animals are respected for their intrinsic value, and where the human-animal bond is strong.

Want to help animals? We have plenty of suggestions. Adopt a pet from a local shelter, join The Humane Society and be a part of our work to help companion animals and wildlife. You will be funding our educational, legislative, investigative, and outreach projects in the U.S. and across the globe.

Or perhaps you'd like to make a memorial donation in honor of a pet, friend, or relative? You can through our Kindred Spirits program. And if you'd like to contribute in a more structured way, our Planned Giving Office has suggestions about estate planning, annuities, and even gifts of stock that avoid capital gains taxes.

Maybe you have land that you would like to preserve as a lasting habitat for wildlife. Our Wildlife Land Trust can help you. Perhaps the land you want to share is a backyard—that's enough. Our Urban Wildlife Sanctuary Program will show you how to create a habitat for your wild neighbors.

So you see, it's easy to help animals. And The HSUS is here to help.

2100 L Street NW • Washington, DC 20037 • 202-452-1100
www.hsus.org

Trademark Disclaimer

Author Acknowledgement

Writing a book is usually considered a lonely process as the writer spends much time alone researching material and writing text. But no writer is an island and during the creative process many people contribute to make a book successful. First, I'd like to thank those friends and family who supported and encouraged my decision to strike out in writing. Particularily I'd like to thank Karen Hipple-Perez, who always had a ready ear to listen to my ideas and to help sort through things; Jennifer Hipple, a fellow writer; my sister, Rosanna Callahan; and my brother, Terry Nelson.

I would also like to thank the participants in my case studies who really made the book with their real life experiences. They were all very open and willing to share their experiences.

Dedication

To my parents, Henry and Suzanne Nelson, who instilled in me a healthy respect and love of all creatures great and small.

Table of Contents

Introduction

Caring for poultry can be a fun and rewarding project. Birds can provide food, eggs, and pleasure, provided you enter into raising poultry with realistic expectations. This book will provide a small-scale poultry farmer with the information needed to raise healthy, productive birds. It will cover in depth how to choose the right breed, how to start a flock on a healthy road, how to use your birds, how to breed birds, and many pearls of wisdom to make sure your birds are as robust as possible.

The United States Department of Agriculture (USDA) defines a farm as "any operation that sells at least $1,000 of agricultural commodities or that would have sold that amount of produce under normal circumstances." Many small-scale farms can easily sell that amount of poultry meat, live birds, or eggs each year to be counted as a farm in official numbers. A typically small-scale farmer will raise fewer than ten beef cattle, or have 50 to 70 sheep or goats, or a small flock of chickens, ducks, or geese. A big point to make is that the average small-scale farmer derives a side-income, not a main income, from the farm.

An important aspect of being counted as an official farm is participation in federal farm programs. These programs vary widely depending upon what product your farm produces and are heavily weighted to support grain farmers. However, livestock and poultry farmers are able to participate in some of these programs from beginning farmer support in terms of loans and technical advice to disaster assistance during declared natural disasters. Each county maintains a Farm Services Agency to assist in determining your eligibility and to guide you in signing up for the programs geared toward your farm operation. You can learn more by following this link to the United States Department of Agriculture's website: **www.fsa.usda.gov**.

A few things have remained constant in agriculture throughout the years. It is a tough business requiring physical labor and work in a field of uncertainties due to weather, disease, and injuries. Market volatility is another big uncertainty. All these factors can combine in a disastrous way to drive a well-meaning farmer from their livelihood, or they can build a person's character and resourcefulness in ways never thought possible.

This book will help you get off on the right foot in establishing a small-scale enterprise. You will learn how to prepare for newly-hatched chickens, ducks, geese, turkeys, quail, guinea fowl, and pheasants. You will learn how to recognize the signs of an ill bird and how to treat a sick animal. A large portion of the book will give you information on the many, many breeds of poultry that are available. *The Appendix will give you a wealth of resources on where to purchase your birds, equipment, coops, and feed.* It will also list organizations and agencies dedicated to providing information and support to poultry owners. In addition, case studies from farmers, poultry experts, and backyard poultry enthusiasts are peppered throughout the book to give you further insight and hints on caring for your own flock. In a nutshell, you will learn the ins and outs of poultry farming, but as with most things, you will learn best by doing. Most likely you will experience a few setbacks during your first year or two, but with practice, research, and determination, your foray into small-scale poultry farming will be a fun and rewarding experience for you and your family.

Before you embark on a small-scale enterprise, you will need to do your research. Research the species and breeds available, the equipment needed, the physical labor requirements, and the markets available to sell the products your poultry produce. To do this, you will have to ask yourself some tough questions: Do you have the resources to finance a poultry enterprise? Is there a reliable source of feed near your farm? Is there an available, near-by market for your product or do you have to create a market? *The Complete Guide to Poultry Breeds* will help you answer these questions and more.

Fresh eggs for sale at a European market.

Part 1

CHICKENS

Congratulations! You have taken the first step and decided to get some chickens for your home use or for a side income. Before you make any purchases, you will want to make sure it is legal for you to keep chickens and to determine if you have adequate room on your property to raise a healthy flock of chickens. You will also want to create a rough budget to see if you can afford to keep chickens. In addition, giving a little thought to why you want to raise chickens will help you develop a plan to market your products.

Having fresh eggs is just one benefit of raising chickens.

Chapter 1

GETTING STARTED: SOME INITIAL CONSIDERATIONS

There are different methods of raising chickens depending on how much land you have available and the zoning laws in your area. Market research may involve investigating where you can sell your poultry meat, eggs, or live birds. Are there any farmers' markets in your area? They typically welcome new vendors, especially those selling fresh eggs. Does the local food coop need a supplier of organically raised meat or eggs? You can grow your own brand locally and build a steady customer base with the help of a coop. If you are planning to raise chickens

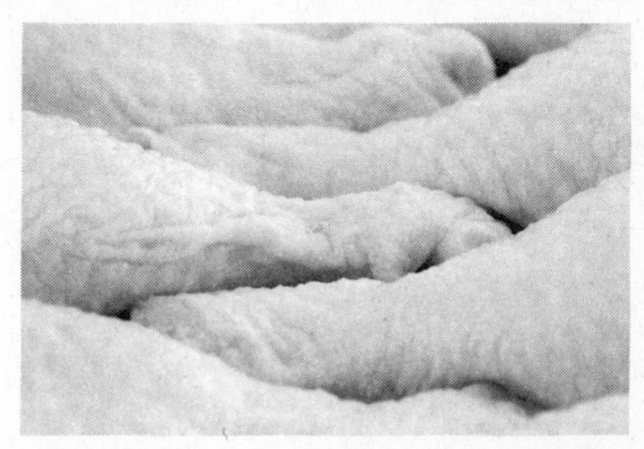

More than ever, consumers want local, organic meat.

for meat you should investigate if there are any local meat processors in your area.

Property: Acreage to Zoning

Finding out if your land is legal for chicken rearing is fairly straightforward. Usually a quick visit or phone call to the county zoning office will let you know if your property is zoned for chickens and how many you can raise. Many municipalities — even large metropolitan areas — allow homeowners to own a few hens. Chickens raised in a city usually must be penned or completely fenced in on their owner's property. They cannot be allowed to roam at will, but roosters are another story. With their loud crowing and implication in cock fighting, they are usually forbidden inside city limits.

This rooster is beautifully colored, but its crow can be a nuisance to neighbors. Keep your neighbors happy with a gift basket of fresh eggs.

The best place to start is to contact your local government office, which is typically housed in the county courthouse. Zoning laws vary according to local regulations, and even among government offices there may be discrepancies. The county clerk, the animal control department, planning board, county commissioners, or the environmental office may all have jurisdiction over the keeping of livestock. When you get information, make sure you get it in writing in case anyone questions you later on. If you are finding that all these offices are telling you "no," you can try to appeal the decision. The law may be an old regulation. Times have changed and you will find that many decision-making agencies may be receptive to keeping a few chickens within city limits.

If you plan to build a shed or coop for your birds you will also need to determine if you need a building permit. Again, a visit to the courthouse and planning board to ask if you need a permit will be in order.

Laws for each county, town, or village will vary greatly so it is not possible to find a complete website listing for each state. Your best course of action is to stop at the local government office and ask them to point you in the right direction. Usually, they will have a wealth of information available and might have information you would not have even thought to ask.

Your neighbors will appreciate if you build attractive housing for your flock.

If you find that your land is zoned for poultry, you will then determine where you will house your flock. A corner of a shed can be used for a pen, but try not to use a shed or outbuilding where you will run engines, as the fumes can hurt the chickens' lungs. Try to have some form of natural lighting and fresh air in the building. A shed or building with windows that open or a door that can be propped open in good weather will make the building chicken-friendly. Alternatively, you can attach a simple chicken run constructed of wood and chicken wire to a

A hen sitting in her hen house.

door or small opening to the building to give your chickens a screened-in porch of their own with plenty of fresh air and sunlight.

If your plans are to eventually let the chickens roam the farmstead, you will have to be careful if you live near a busy road. Chickens may wander in search of grit or bugs and can easily become accident victims. Instead, you might want to consider constructing a portable pen enclosed with chicken wire to allow your chickens more room to peck and exercise. This will also protect the chickens from their enemies; predators such as rats, cats, dogs, raccoons, owls, hawks, and coyotes all prey on chickens and their eggs.

Pocketbook: Start-up Costs and Feed Bills

While chickens are not as expensive to keep and feed as most farm livestock, you will want to track your expenses to make sure you are not going over your budget. If you are starting from scratch, the following is a list of some items you will need to purchase for your chicks. Keep the receipts for your purchases, as most agricultural items can be tax deductible, provided you meet the IRS definition of a farm.

- One 1-gallon water device per 50 chicks.
- A feeder that provides 1 inch of space per chick.
- One heat lamp (250-watt bulb) per 25 chicks, or one brooder. Sizes vary, so you will need to check manufacturer's instructions as to how many chicks can be brooded per brooder.
- One to two bags of shavings or sawdust. You will use half of a 3-cubic bag per week for 50 chicks.

- An electrolyte powder picked up at the farm store to mix in chicks' water during the first few days. This provides the chicks with extra nutrients.

- A commercially prepared bag of chick starter.

Farm stores or farm catalogs are good sources of this necessary equipment. Another option is to scour farm auction sales bills to see if any poultry equipment will be on the auction block. If you are particularly handy, some equipment can also be built from scratch. Free Chicken Coop Plans (**www.freechickencoopplans.com**) has a great tutorial on how to build inexpensive chicken feeders and waterers.

Prices for chicken coops will vary. A simple coop providing room for 25 hens will cost around $100 to build if you do all the work yourself, but if you need to purchase a chicken coop they can be expensive: Near $1,200 will get you a professionally made coop that houses ten to 20 hens that will last a lifetime with routine maintenance. A small garden shed can also be converted to a chicken coop as well. Of course, if you have existing buildings that are suitable for chickens, your start-up cost for a building will be negligible. Your prior research into zoning laws will help you determine any permits you will need if you are building a permanent pen for your chickens.

FeatherSite (**www.feathersite.com**) offers links to free or inexpensive chicken coops. Another inexpensive design using PVC pipes and chicken wire can be found at PVCPlans.com (**www.pvcplans.com**). This plan can be downloaded and printed and within a few hours of work you will have a lightweight, portable pen for your birds.

Further down the road into chicken farming, you will find you will need to purchase supplies to butcher your chickens or to keep your meat and eggs fresh. If you are keeping chickens only for home use, you will probably have most of the supplies right in your kitchen: knives, large pots to heat water, and a refrigerator/freezer. However, if you plan on selling poultry products in large quantities, you will want to invest in a separate refrigerator/freezer and other butchering and storage supplies like plastic freezer bags, egg cartons, and perhaps a chicken plucker, which is a machine that pulls out the hairs and feathers of a chicken. eNasco, a leading provider of agriculture materials, carries all these supplies (**www.enasco.com/farmandranch**).

After you make the initial investment in equipment and housing, your main ongoing cost will be feed. An adult chicken will eat 4 to 6 ounces of feed a day, while young chicks less than 3 weeks of age will eat less, and the heavier breeds of chickens will eat more. This rough estimate will help you determine how much feed your chickens will eat. A 50-pound bag of chicken feed weighs 800 ounces. If you are feeding ten chickens each 5 ounces of feed a day, the bag will last 16 days, and if the bag is priced at $12, you will spend approximately $24 a month on feed. Of course, as chickens get older they can derive more of their diet from bugs, vegetable scraps, and less-expensive feed than those chickens in the starter and growing stage. Do not be tempted to short-change young, growing chickens by feeding them a poor-quality diet. This will adversely affect their health, giving you weak and nonproductive adult birds. *See Chapter 3 of this book to find out the nutritional needs of chickens at each stage of growth.*

You need cartons if you plan to sell your eggs.

One caveat: Feed prices vary widely depending upon the current price of grain and depending upon the region of the country you live in. As an example, in 2007, corn was priced as high as $13 a bushel. Now, the average 2010/11 farm price is estimated to be $3.20 to $3.80. Such wide fluctuations are not uncommon due to weather conditions, grain demand by manufacturing companies and livestock feeders, and other market conditions. So what is valid today for feed prices may not be the same six or even three months from when you calculate costs. That is just the way the market fluctuates and why it is so important to track your feed expenses so you are not priced out of business.

To help with the costs of operating a poultry farm and to help to cope with varying prices, there are some available grants and loans to people planning a larger, small-scale chicken farm, which is a farm raising anywhere from 100 to 20,000 chickens. Sustainable-farm organizations are often good sources for these grants and loans to help enhance the local rural economy by supporting small farmers. The Sustainable Agriculture Research and Education organization (**www.sare.org**) provides competitive grants to farmers and ranchers in the United States, and many state agricultural departments also provide similar programs. The National Sustainable Agriculture Information Service (**www.attra.ncat.org**) offers a comprehensive, state-by-state list of sustainable farm agencies, grassroots groups, and nonprofits.

The Rundown of Raising Chickens

The scientific name for the domestic chicken is *Gallus domesticus*. They come in different types: show or ornamental breeds, bantams, layers, meat-type, and dual-purpose. Which type you choose for your farm depends upon your ultimate purpose for raising chickens. The show, ornamental, and bantams are primarily raised for breeding and showing at fairs and competitions. Before you decide to raise chickens, consider a few points. You will need to decide the primary purpose your chickens will serve. Do you want a fresh supply of eggs for your family? If so, keep in mind an average hen will lay 260 eggs a year – about five per week – for one to two years. Do you want to raise eggs to sell to the public? Then you will need to conduct some market research on your potential customers' needs and the demand for farm-fresh eggs in your area.

It is important to know that class, breed, and variety are all different aspects of defining a chicken. A class usually relates to where the chicken originates. Many breeds will be a part of a class; for example, the American class. A breed is defined by the characteristics common for that type of chicken. These characteristics include meat characteristics, average weight, bone structure, and habits. A variety of chicken will be a variation that occurs in a certain branch of a specific breed. An example would be different feather coloring or a unique characteristic. Do not get too hung up on these words; it is more important to understand what traits you want your flock to have and select your breed according to your specifications.

This black Pekin Birchen cochin is a bantam hen.

This lemon Pyle rooster is a special, rare breed.

Chickens are often categorized by their purpose:

Layers: These are specially bred to be prolific egg-layers but are skimpy meat producers.

Meat birds: Meat-type birds have been bred to gain weight quickly on the least amount of feed, but the hens cannot be expected to produce a large quantity of eggs. However, the eggs from a meat-type hen are perfectly fine to eat.

Dual-purpose: These breeds are multi-purposed, combining the qualities of a meat-type and layer bird. However, they do not gain weight as quickly as the meat-bird, nor do they lay eggs as heavily as the layer breeds. Many small-scale farmers rely on these dual-purpose chickens to supply their family and customers' needs.

Ornamental: These chickens are birds of a different feather. They are raised primarily for their beautiful or unique plumage – basically, these chickens are pretty to look at. Chicken fanciers will proudly enter their best birds in poultry shows just for bragging rights.

Chickens can be additionally categorized as pure breeds, hybrids, and bantams. Pure breeds are chickens bred from members of a recognized breed, strain, or kind of chicken without the introduction of other breeds over many generations. Hybrids are a mix of different breeds, often bred to accentuate or eliminate certain qualities of a particular breed. Large poultry companies that specialize in either meat or egg production typically breed hybrids. Bantams are mini chickens and can be any breed, but they are typically ¼ to ½ the size of the average bird.

A Colombian cochin chicken resting on a stone wall.

To find out if there is a demand for eggs in your area, begin by approaching your friends, acquaintances, or neighbors to see if they are interested in purchasing eggs on a consistent basis. If you want to expand beyond selling a few dozen eggs a week, a local farmers' market or the local grocery store may be interested in purchasing your eggs; you can also become a vendor at a local farmers' market. Each market is governed by its own rules and will probably charge a minimal fee for your booth or table. This is also a great way to meet other like-minded people and trade hints and tips for a successful poultry enterprise.

Six hens should adequately supply enough eggs for eating and baking for a family of four. Egg production will drop a little in the winter months, as egg production is related to the length of daylight. A chicken relies upon a certain amount of daily light (approximately 14 hours) to stimulate its reproductive system. If you want to try to keep up egg production during the winter months, keep the hens in an insulated building with a light burning during the evening to add a few extra hours of light.

Do you primarily want meat from your chickens? A broiler, or male meat bird, will be able to be butchered at around 6 to 8 weeks old, yielding 4 to 5 pounds of chicken meat. Will you have the freezer space to store your butchered chickens? Do you plan to perform the butchering process yourself or is there a nearby butcher shop that will butcher small batches of chickens? These questions will be answered throughout this book.

Fresh, organic eggs are a healthy addition to your diet.

If you plan to raise ornamental or show birds, do you have a well-insulated barn to house them? These birds need a bit more pampering to keep their feathers at their best and to keep them from becoming soiled. In addition, in cold-weather areas, chicken combs are prone to freezing. A show bird with a frostbitten comb or wattle will be frowned upon in the show ring.

You may decide you want to breed chickens to sell as live birds. Your chickens can have a desirable quality such as unique size, egg color, or feather color that may be in demand. Chickens are easy to breed and reproduce fairly quickly, and your small flock may soon become large. Never breed sick or deformed birds — health problems and deformities can be a genetic problem. Only your healthiest birds should be kept for breeding.

The same applies to selling your birds. Only healthy birds should be sold, and you should check with your local extension service or state board of animal health before selling birds to see if there are any restrictions on the sale of live birds. Chickens can be sold word-of-mouth, through classified ads in the local newspapers, at farmer markets, or on websites such as Craigslist (**www.craigslist.org**) or eBay (**www.ebay.com**).

These roosters are for sale at a market.

These questions and issues are just the beginning when deciding to establish a flock of chickens, even if it is just on the small scale. By reading the rest of this chapter, conducting your own research, and evaluating your needs for housing and zoning restrictions, you will be able to decide if chicken-raising is for you and start this fun project with reasonable expectations.

Housing your chickens

By far, the most common way a small flock of chickens is raised is in a pen, commonly called a chicken coop. Because chickens are vulnerable to predators, many people have found it much safer to keep their hens and broilers penned inside a predator-proof building. Chickens are usually fine with this; however, if your pen is overcrowded, the chickens may have trouble with respiratory disease due to build-up of dust and fumes from the manure. To combat this problem, provide access to an enclosed, outside area to allow the chickens to enjoy the fresh air and to catch any bugs that stray into the enclosure. A variation of a traditional chicken coop can be found at Omlet (**www.omlet.us**). This unique coop, dubbed Eglu, is shaped like a cone or comes in a cube shape. The coop where the chickens sleep and lay eggs is on one end, and a run, or pen, is attached to the other. These coops can be used in areas as small as 20 feet by 30 feet and can house two to four chickens. They are portable and can be easily cleaned.

Some people swear by the free-range method of chicken rearing: After the hens or broilers are fully feathered-out (at around 5 to 6 weeks of age), the chickens are released from their brooding area and allowed to roam the farmstead at will, but at nighttime, most people will let their chickens come home to roost in a secure building. This is by far the best practice as you can never tell if a neighbor's dog, cat, or a wild animal will use the dark night to help themselves to a midnight snack of fresh chicken. Many breeds of domestic chickens, especially the heavier breeds, have lost their ability to fly; therefore, they do not have an easy escape plan to flee from their enemies.

This Scots dumpy lives on a free-range farm.

One interesting methodology for raising chickens is to actually bring the chicken pen to a pasturing area. The farmer will use a portable chicken coop, which has a handle and wheels to move the entire pen by hand or is built with skids so it can be moved with an all-terrain vehicle (ATV) or small tractor. The coop is designed with an enclosed area so the chickens can get out of rainy weather and lay their eggs, and attached is a large pen made of chicken wire so the chickens can eat grass and bugs. Orchardists find this a handy way to control grass and insects among fruit trees while ensuring the safety of the chickens.

Processing your products

These free-range chickens forage through grass for bugs.

Soon you will find that your chickens will be ready to be used as a meat source, for eggs, or to sell as live birds. Before you sell your first product, you will need to understand the "rules" surrounding the sale of poultry and their products. These laws and regulations are in place for the consumer of the poultry products and to protect the poultry industry as a whole against diseases.

Poultry has gotten a bad rap as a carrier of many harmful food-borne diseases. Salmonella is frequently found in eggs and fresh meat, which can also become contaminated with Campylobacter. This is why it is not recommended to eat batter made with unpasteurized eggs. It is also why you need to bleach cutting boards used to prepare raw poultry meat. While you might not be able to prevent your birds from being a carrier of these disease organisms, you can prevent contaminating the eggs by providing nest boxes with clean bedding material and by properly butchering your birds.

Also, if you sell live birds you should check with your local veterinarian's office or state animal board to determine rules for testing live birds for diseases prior to selling. By doing so you can offer your buyer an additional service by selling disease-free birds while also doing your local poultry farmers a favor by making sure only healthy birds leave your farm.

Take every precaution you can against salmonella.

If your plans only involve using your birds for meat for yourself or family, will you be able to stomach the task of butchering? It is a fairly straightforward process. *Chapter 5 will walk you through the process.* Another option is to hire a processor to do the work of butchering and packaging for you. While prices vary, a typical processor will charge up to $3 a bird for their services. However, it might not be as simple as you think to find someone to do the butchering because local meat shops do not necessarily want to invest the time and manpower involved in butchering. This is primarily because the feathers need to be picked off the bird, which is a somewhat time-consuming task. Many people will find they need to haul their live birds quite a long distance to find a processor. A call to your local farm extension service may help you with this. The National Sustainable Agriculture Information Service provides a state-by-state online database of small poultry plants and services at: **www.attra.ncat.org/index.php/poultry_processors**.

Each state has its own rules regarding the sale of meat from farms. This is basically due to food safety issues, as chicken meat is notorious for harboring dangerous bacteria. During the butchering process, it is almost inevitable that fecal or digestive tract material will contaminate the surface of the skin or meat if you do not carefully remove the digestive tract. If you use the meat strictly for yourself or your family's use you do not have to worry about state laws, although you will want to keep the butchering process as clean as possible by using clean, cold running water to rinse and cool the birds. The same concerns surround the sale of eggs and live birds off the farm. Again, these are in place to protect the consumer from foodborne illness. Chickens with Salmonella infections lay contaminated eggs. In addition to foodborne illness, with selling live birds there are concerns surrounding the spread of disease to other poultry.

If you plan on entering poultry sales, do some sleuthing prior to selling your products. A good first step is to consult your local farm extension agent for materials specific for your state. You can also check with your state board of animal health for the ever-changing food safety rules and regulations. States have jurisdiction over the sales of meat and poultry within the state. If you plan to sell your products across state lines, then this falls under federal government jurisdiction. The National Association of State Departments of Agriculture (**www.nasda.org**) has a listing of each state's department of agriculture, contact information and a link to the State's official website. Visit the United States Department of Agriculture Food Safety and Inspection Service (**www.fsis.usda.gov**) to find information on interstate sales of poultry meat and products.

A History of the Chicken and Chicken Production

The modern chicken has its roots in the jungle of Southeast Asia. While Charles Darwin thought all chickens descended from the red jungle fowl, it is now known that the gray jungle fowl also contributed to the gene pool of the domestic chicken. From the breeding of these two wild fowl, we now have hundreds of breeds, types, and strains of chickens.

Scientists are not in agreement as to when the chicken was first domesticated. Thailand may hold the distinction of first domesticating the chicken, but many believe there may have been multiple areas in Southeast Asia where wild jungle fowl were captured and kept for egg production. The earliest archeological evidence of domestication has been found in China dating back to 5400 B.C. The red jungle fowl gave us the white-skinned chicken, while the gray jungle fowl gave us a yellow-skinned chicken. Carotenoids, a natural, fat-soluble pigment found in the chickens' feed, form the yellow pigment in their skin. An enzyme coded by genes breaks down the carotenoids and releases vitamin A. Chickens with this gene that eat high levels of carotenoids develop yellow skin and legs.

Having chickens in the backyard used to be much more common than it is now.

Prior to the early 1900s, poultry was raised in the backyards and farmyards for home use and sold for extra money for the household. Chickens were not specifically bred or fed for meat production; instead, old hens or roosters would be occasionally plucked from the flock for Sunday dinner.

In 1923, Celia Steele, a housewife in Sussex County, Delaware, purchased 500 chicks intended to be sold for meat. She had the notion that chickens could be raised for more than eggs, which is why they were typically used prior to this time period. Steele had the foresight to see that chickens could be sold as broiler chickens and not just egg-layers. Steele's first flock sold for 62 cents per pound, which is the equivalent today of close to $15 per pound. Housewives and restaurant owners discovered the versatility of preparing chicken (frying, broiling, as stew meat, and roasting), causing demand for chickens to increase.

By 1926, Steele's flock increased to 10,000 as she discovered that vitamin-enhanced feed could produce a heartier bird. Less than ten years later, the Steeles prospered to own seven farms. Even today, Sussex County, Delaware is the top broiler-producing county in the country — delivering millions of chickens each year. It was not until the 1940s that modern poultry production really took off with the advent of integration of the chicken industry.

A large chicken house housing many chickens.

According to the National Chicken Council (**www.nationalchickencouncil.com**), feed mills, farms, processing operations, and hatcheries were interdependent on each other prior to the 1940s. The feed mills would loan money to the farms to buy chicks from the hatcheries, and when the farmers sold the flock to the processors, they would use the money they received from the processors to pay back the feed mills. This practice became more common and regulated as chicken consumption increased. Refrigeration also helped the industry because it allowed consumers to store their meat for longer. Factory farming produced more products for less money and raising chickens that scratched around in the backyard became less popular and not as lucrative.

In the 1950s, production increased to meet the baby boomers' needs. Vertical integration, the act of one company controlling all processes from marketing to production to reduce costs, helped manufacturers afford new technology offered at the time, which increased sales and profits. Entrepreneurs with vertical integration systems controlled most of the chicken industry at this time. In the 1960s, marketing expanded to television and print, making poultry brand names more recognizable and more popular than ever.

Automation technologies of the 1970s helped producers meet consumer demands. Regulations and laws became more focused on production as people became more educated on the poultry's nutritional values, diseases associated with chickens, and the process of speeding up chicken growth. The government and the public scrutinized the cleanliness of the chicken plants, the environments the chickens lived in, and the way the birds were killed as both groups became more aware of the way the chicken industry operated.

The growing chicken market

During this time, poultry was not the only industry with stricter regulations; overall, the United States was setting higher standards and fine-tuning its food markets. Their eyes were open to the potential harm of unsafe practices, and people closely monitored the progress of food production. Demand was steadily increasing, and chicken producers enhanced chicken growth to meet these needs because faster-growing birds meant there was more poultry available in a shorter amount of time — this increased profits.

In the 1980s, demand for poultry expanded further when fast food restaurants added chicken tenders and nuggets to their menus. McDonald's, now a worldwide fast-food brand, introduced Chicken McNuggets to the nation in 1983, and within a month, it was the second largest chicken retailer in the world. This chicken sensation helped increase poultry sales. In 2003, the amount of chicken nuggets sold in restaurants overall increased to more than 200 percent from 1990s. McDonald's is credited with introducing the nugget into the American way of life. It was not just a fad; the nugget became a staple in the American diet, a staple that appealed to all age groups.

By the early 1990s, chicken sales surpassed the sales of beef. Statistics from the National Chicken Council state that during this time period, growing amounts of chicken produced in the United States was exported to other countries, including the Soviet Union. In 2001, exportation of poultry from the United States to other countries reached an all-time high. Not only were poultry broilers booming within America, they were also increasing globally.

This large chicken farm produces eggs and meat.

Stricter laws have developed over the past six decades to ensure the safety of the birds produced for consumption, and the USDA enforces these rules. These laws became necessary when the way the animals were treated was deemed inhumane and factory conditions were ruled unclean. Because of these new rules, birds are less expensive than they once were, have more meat on their bones, and are produced in cleaner, safer environments. Although there is still debate on the humane treatment of the animals, these regulations aim to achieve the best possible conditions for both the workers and the birds.

About 10 billion chickens are hatched in the United States each year. Of these 10 billion, 95 percent of chickens are raised in large barns on a small tract of land. The farmers that grow these chickens contract with the large chicken processing firms – Tyson and Perdue Farms control most of the poultry farms – that have a set schedule as to how many chickens each farm can grow, the growing conditions, the feeds that are provided to the chickens, and at what stage and when the chickens can be processed. In most cases, these processing firms own the chickens from birth date to slaughter date, and the farmer owns the buildings and some equipment, which for the most part is bought with money loaned from the company. There are few independent growers, but the most common source of chicken meat and eggs are by far the large commercial farms controlled by a few large processing firms.

Large, modern farms feed their poultry with automatic feeders.

Each firm will have their own commercial breed of chicken to ensure uniformity of size and eggs. The firm will typically own their own breeding flocks and hatcheries. Most meat-type commercial birds are crosses with the Cornish breed of chicken, while the layers are Leghorn chicken crosses. These breeds have consistently produced fast-growing birds that consume the least amount of feed while growing or laying.

However, hand-in-hand with this great productivity comes some health problems. Fast-growing broilers are frequently victims of cardiac arrest and leg problems because their muscles often grow faster than what their immune system can support. The layers can be nervous and prone to flightiness — an undesirable trait in a small flock. No one likes to have birds that stir up dust and flee from people at the least provocation.

As a small poultry flock owner, you can overcome some of the negative stereotypes associated with the large-scale chicken farms. One interesting movement of late is the local foods movement. People are searching for a more humane, more sustainable food source. You can tap into this desire by promoting your birds or eggs as a humanely-raised bird with a more comfortable existence. Remember, you can only claim to be more humane than large-scale farms if you put into practice the tenets of allowing your flock room to roam and live as naturally as possible — while still protecting your investment.

Consumers appreciate chicken that is raised on pasture.

When selling your products, you will find non-farm people do have an interest in supporting smaller farmers through purchasing meat and eggs from their farmer friends or neighbors. Often, people will want to support a local farmer simply to support the economy of their own community. Associating your image with local business makes a great marketing tool, especially if you are able to tell your buyers and supporters specifics about the raising of the birds. For example, sharing information about the diet of the birds and how they interact with each other and their human providers will solidify your local, small-farm image. By doing a little promoting of your chickens and your farm, you can steadily build a loyal following of buyers — which leads to a profit on your investment.

CASE STUDY: FROM CORPORATE TO COOP

Sandy Gilbertson
Armstrong Acres
Backus, MN

The Gilbertsons planned to begin farming a small plot of land they inherited in 1990 someday. In the summer of 2009, that day finally came.

"I had a job in a large corporate office; seemed like a waste of every day," said Sandy Gilberston.

He and his wife chose to organically raise chickens for their fresh eggs and meat. They started off with ten laying hens and 25 meat chickens. They now sell some of their products at a farmers market and will soon have an in-home store as well.

"I feel raising them without confinement and with no unnecessary medications is the way to go," he said.

Since they have never had a health problem with their chickens, they must be doing something right. In fact, their biggest concern is the odd loose dog that sometimes claims a chicken.

Gilbertson enjoys his new endeavor very much. He dedicates about 20 hours a week to the birds, and he finds them a more enjoyable career than life in the corporate world.

"They are easy and very entertaining," he said

Children, he said, would also enjoy his "pets."

"They are quite spoiled and would be great with children."

Chapter 2

CHICKEN BREEDS TO CHOOSE FROM

Chickens are categorized as pure breeds, hybrids (egg-laying and mixed), and bantams. Pure breeds are chickens bred from members of a recognized breed, strain, or kind of chicken without a mixture of other breeds over many generations. Hybrids are of a mixed lineage, often bred to accentuate or eliminate certain qualities of a particular breed. Bantams are mini-chickens and can be any breed, but they are typically ¼ to ½ the size of the average bird. Bantams make great pets, especially in residential neighborhoods, because they are petite and require less space. They have the same characteristics as their full-size counterparts, but are just a smaller size.

These two hens have beautiful, black and white feathers.

One of the most amazing things about chickens is that there are more than 500 chicken breeds throughout the world. In some countries, chickens have a place in the national spotlight. Japan holds great pride in their Onagadori ("honorable fowl" in Japanese) chicken, a bird with tail feathers up to 40 feet long. Egypt has the scrappy Fayoumi, a breed of chicken believed to be centuries old. These birds are not often found outside of Egypt, except as a show-breed. Japanese fishermen keep Ayam Bekisar roosters aboard fishing boats for their loud crows as a means to communicate with other boats.

With all the amazing varieties and breeds of chickens, no one book can give adequate coverage to each and every breed; however, this book will present a wide variety of chickens that are accepted in the American Standard of Perfection – the official chicken breed standard in North America – to give you a strong background in the chicken breeds out there. To give some sort of order to the breeds, they are presented under four categories: meat breeds, laying breeds, dual-purpose breeds, and ornamental breeds. Within the breed categories there will be much variation in color, size, and production, and ultimately, it will be up to you to decide which breed will best fit your needs. You may even find yourself purchasing a few different breeds, as it can be hard to settle on just one. For more information on chicken breeds here are some good websites:

- Ithaca College's Henderson's Chicken Breed Chart: **www.ithaca.edu/staff/jhenderson/chooks/chooks.html**
- My Pet Chicken: **www.mypetchicken.com/chicken-breeds/breed-list.aspx**
- FeatherSite: **www.feathersite.com/Poultry/BRKPoultryPage.html#Chickens**

- Poultry Pages: **www.poultrypages.com/chicken-breeds.html**

Meat Breeds

The meat-type chicken is not necessarily considered a breed category, but rather strains of breeds that have been specifically developed for certain traits. These traits include rapid growth, larger yield of breast muscle, or efficiency at converting feed to muscle. Typically, each large broiler-processing firm will have its own specific breeding program to obtain the traits it desires for its production lines. Popular breeds to use in commercial strains include Cornish and Plymouth Rock.

This silver Dorking rooster looks alert in the field.

Dorking

Rooster: 10 to 14 pounds
Hen: 8 to 10 pounds
Bantam: up to 3 pounds

This breed is believed to have originated in Rome about 2,000 years ago, and it was further developed in Britain. Their specific purpose is to be a plump, meaty bird with enough meat to feed a family. Its colors are red, silver-gray, and white, and it has white skin. Their legs are clean, short, and white with five toes. Their earlobes are red, and their eggs are white. Dorking hens will lay about 100 eggs in a season. These birds are docile and easy to handle. They do not like confinement and will be happier free-range. You will need lots of space for them to forage for insects and wander. Dorkings take up to two years to mature, and they live up to seven years. Feed that is balanced and provides proper nutrients will give you a meatier bird.

Langshan

Rooster: 9 ½ pounds
Hen: 7 ½ pounds
Bantam: 2 to 3 pounds

The Langshan chicken originated in China and was discovered in the 1800s. They were popular in the late 19th century in America because of their qualities: They can tolerate all climates, are meaty, and are adequate egg producers. They also have a gentle temperament. They are blue, black, and white, with a single comb and white skin. The blue and black birds have a green or purple sheen to them in sunlight. Green is a more desirable color and represents a more prestigious bloodline. Their eyes are dark or black. Langshans have long, clean legs with four toes, and some varieties have lightly feathered legs. By today's standards, these birds are big but not as meaty as some commercially produced chickens. These chickens stand tall, about 20 to 24 inches high, and their tails can be long, are erect, about 18 to 24 inches, and are carried at a high angle. This fowl are medium egg producers, providing 140 to 150 eggs yearly. Their eggs are light to dark brown, and in some cases a bird produces plum-colored eggs. Langshans are a good choice for new chicken owners because the birds are gentle and docile.

Cornish

Rooster: 10 ½ pounds
Hen: 8 pounds
Bantam: 3 to 5 pounds

Cornish chickens were developed primarily for meat in Cornwall, England. The breed has contributed the most to the development of the broiler industry because of their muscular bodies and excellent carcass shape. Their feathers are short and close to the body, and they have yellow skin with feathers ranging from white to buff and white-laced red. Cornish chickens have yellow skin. Their eyes are reddish brown, and they have a single pea comb. Their legs are clean, and they have four toes. These birds are not proficient egg-layers, but they do produce brown eggs. They are not friendly poultry so they are not the best choice for a pet. Cornish chickens tend to be noisy and are energetic birds that are always on the move. They do well in confinement, but they need exercise to help them to keep their muscular, meaty shape. Cornish chickens are crossed with Plymouth Rocks to give a commercial strain (Cornish-Rock) of meat-production bird.

Brahma

Rooster: 10 to 12 pounds
Hen: 7 to 9 pounds
Bantam: 32 to 38 ounces

The name of this bird comes from the Brahmaputra River in Asia. Brahmas are sometimes referred to as the "King of Chickens" because of their large size. They come in an assortment of colors, including buff Colombian (meaning they have black tails or black-tipped feathers on their tail), gold, and white. Their coats can be light or dark in color. They have red eyes and a small, single pea comb, and their legs are feathered. Brahma roosters are docile, even somewhat submissive compared to other breeds. Brahmas are fairly good egg-layers and produce approximately 140 brown eggs yearly. Although they are large birds, they are gentle and easy to handle. They take up to two years to mature. Brahmas need a dry environment, but they can fare well in hot or cold climates. They do not fly and are content behind a 2-foot fence.

Faverolle

Rooster: 9 to 11 pounds
Hen: 7 ½ to 9 ½ pounds
Bantam: 2 pounds

This breed of chicken is a cross-breed developed originally to produce hearty, plump chickens. The lineage of the Faverolle is most likely a mixture of several breeds including Houdan, Dorking, Maline, white-skinned Brahma, and the common five-toed fowl. The colors are black, buff, laced blue, salmon, white, and ermine (a light-colored bird). They have light red eyes and a single comb. Their legs are lightly feathered, and they have five toes. They are productive egg-layers and lay about 100 light brown eggs each year. Faverolles will lay eggs throughout the winter months. These are active birds that are always on the go and need room to roam. They are gentle and sometimes can be bullied by more aggressive breeds, such as the Cornish, old English game, or modern game chickens.

Cubalaya

Rooster: 6 to 7 pounds
Hen: 4 to 5 pounds
Bantam: 3 pounds

This hearty bird comes in white, black, black-breasted red, or blue wheaton (having a dark blue body with a rusty coat on top). They are good egg-layers, their eggs are cream-colored and tinted but they are raised primarily for their meat. They are a beautiful ornamental bird with long tail feathers that curve downward, which is known as a lobster tail. They have reddish-brown eyes, red wattles, and a pea comb. They are fairly rare in the United States, as they originated in Cuba. This breed is friendly and can be trained to eat out of your hand. They can endure any climate and do well in confinement.

Laying Breeds

The egg-laying breeds of chickens are primarily descended from birds that the American Standard of Perfection terms the Mediterranean Class of chickens. The birds usually have small bodies not suitable for meat production. Instead of putting their feed into building muscle mass, the nutrients are put into egg production. Commercial strains of egg-layers are either derived from the Leghorn breed (giving us white eggs) or the Rhode Island red breed (giving us brown eggs). Each hatchery, a place where eggs are hatched artificially, will have their own breeding program and name for their commercial layers.

Ancona

Rooster: 6 to 7 pounds
Hen: 4.5 pounds
Bantam: 22 to 26 ounces
Eggs: 260 per year

Originated near the port city of Ancona in Italy, this breed is a mottled black with white-tipped feathers. The original breed from Italy was colored red, brown, and white. The feathers can also have a green or purple tint. The comb is single and lopped, and it has long wattles. The breed standard calls for a red face that is free of feathers and yellow legs with black mottling. The hens lay small to medium eggs that are white, but they do not like to sit on their eggs. This breed can be flighty and tend to avoid human contact. However, if they are handled when young they can become docile and will follow you around looking for treats. They are excellent foragers for food. They are active, alert birds, and these traits, combined with their darker color, help them avoid predators while they are foraging. The breed also does well in colder climates.

Araucana

Rooster: 5 pounds
Hen: 4 pounds
Bantam: 26 to 28 ounces
Eggs: 200 per year

The Araucana is a hybrid of two South American chickens: the Quetro and the Collonca. They were first bred by the Araucanian Indians of Chile, from which they derived their names. Feather coloring in the Araucana come in almost every color and hue: partridge (black stripes that meet at the middle of the feather then move outward), silver-blue partridge, yellow partridge, fawn, wheaten (creamy tans), white, black, and lavender. Araucanas can be tailed or rumpless (without a tail). They have a pea comb that is low to the head with three ridges. They have a clean leg without any feathers. A unique feature of these birds is the tufts of feathers they have by their ears. The gene that codes for the ear tufts is also lethal in chicks carrying two copies of this gene. Araucanas are a good choice for novice bird owners because of their smaller size and docile temperaments. They do have high energy and enjoy foraging for food. The hen produces blue-green or turquoise eggs. Some varieties of this chicken also lay pink eggs or a brown egg with pink hues. The eggs from this bird are perfect for Easter decorating, gifts for neighbors, or just to add some variety in your egg carton.

Sussex

Rooster: 9 pounds
Hen: 7 pounds
Bantam: 4 pounds
Eggs: 240 to 260 per year

These birds are also known as speckled Sussex in America, and originated in Sussex County, England. They are a plump bird, available in white, silver, red, brown buff, and speckled. Avoid exposing this bird to excessive sunlight because their coat has a tendency to become brassy. The speckled variety gets more speckled with each molting, making it an attractive bird. Sussex chickens have a single comb and a clean leg. The earlobes and eyes are both red, and their skin is white. Their eggs are large and are a cream to light brown color. Because they are alert, curious, and docile, these chickens also make great pets. They can be free-range or penned. This is a hearty breed that will lay eggs even in the coldest of weather, a trait not typically seen in all breeds.

Leghorn

Rooster: 7 ½ pounds
Hen: 5 ½ pounds
Bantam: 1 pound
Eggs: up to 270 per year

The Leghorn has contributed to the development of most egg-laying strains of chickens. They are good foragers and enjoy a chance to roam in grass. In America, the colors of the Leghorn range from white, black, red, Columbian (mostly white body with a black tail or black wing tips on tail), partridge (black stripes meeting at the middle of the feather then moving outward), brown, silver partridge, and black-tailed red with white skin. Their combs can be single or rose, meaning it is almost flat on top and fleshy with small round protuberances. In hens, the comb will typically flop to one side. They have clean legs without feathers and red eyes. The Leghorn also has white earlobes. Leghorn roosters are somewhat aggressive, and the breed in general can be excitable and noisy. Leghorns are shy around humans and are flighty birds. Ideally, they need large, tall coops that allow movement but are secure. It is also a good idea to have some trees with branches for the birds to perch on, as this will also help satisfy their desire to fly. The bantams are calmer than their larger counterparts.

Australorp

Rooster: 10 pounds
Hen: 8 pounds
Bantam: 28 to 36 ounces
Eggs: 200 per year

These chickens have white skin and feathers that are black and blue with a beetle green sheen that is enhanced by the sunlight. Their comb is single and is bright red. The eyes, beak, and earlobes are dark. Their legs are clean and are a slate blue, except for the toes and soles of the feet that are white. Their temperament is quiet and gentle, making them perfect for children to handle, and the neighbors will not even know they are there. They produce brown eggs. Australorps do not fly because they are too heavy. Their home and run do not have to be especially secure to keep them in, just secure enough to keep predators out.

Catalana

Rooster: 6 to 7 pounds
Hen: 4 to 6 pounds
Bantam: 3 pounds
Eggs: 150 to 225 per year

This bird is popular in South America and Spain but is also available in North America. They are hardy birds that are good egg-layers. They are buff in color, and some have black tails. They have a single comb with six points, and the male's comb stands erect. They also have a clean leg and four toes. Their wattles are red and their earlobes are white. The hens lay medium-size eggs, which are creamy colored or sometimes tinted pink. Although these birds are not aggressive, they are active and flighty. They do not particularly like confinement and would prefer free-range living with lots of room to move. The males can be aggressive toward each other. This breed fares well in the heat.

Dual-Purpose Breeds

As the name implies, the dual-purpose breeds will give you hens that are good egg-layers and broilers that will produce a meaty bird.

They will not give as many eggs as the egg-laying breeds nor as much meat as the meat-type chicken, but for most small flock owners they will produce enough eggs and meat for family use.

Andalusian

Rooster: 7 pounds
Hen: 5 ½ pounds
Bantam: 24 to 28 ounces
Eggs: 150 to 180 per year

This breed had its start in Spain and was further developed in the United States and England. The Andalusian lays white eggs. The feather colors seen in the breed are blue (which is the required color to show this breed of chickens), black, white, or black and white. The adult blue chickens will have slate-blue feathers with a narrow ridge of dark blue. This breed is an active forager, keeping feed costs down during warm weather when the chicken can remain outside. However, the bird is so active that it can run fast, making capturing one a difficult experience.

Delaware

Rooster: 9 pounds
Hen: 6 pounds
Bantam: 30 to 34 ounces
Eggs: 180 to 200 per year

This breed was developed in the United States. During the 1940s, this breed was popular for meat production, but by the early 1950s other breeds replaced the Delaware for use in meat production. The Delaware breed's feathers are white with black markings commonly present around the neck, wing tips, and tail. They have a white comb with five points. It is a heavy, dual-purpose chicken laying extra-large, brown eggs. The Delaware is a great forager and has a calm disposition. The breed's conservation status was listed as "critical" by American Livestock Breeds Conservancy in 2008, but this status was moved to "threatened" due to conservation efforts.

New Hampshire red

Rooster: 8 ½ pounds
Hen: 6 ½ pounds
Bantam: 5 ½ pounds
Eggs: 150 to 180 per year

This is a dual-purpose breed, which makes a good meat bird. It also lays brown eggs and has beautiful red-brown feathers. The hens will grow broody and they make good mothers. They are active foragers. The New Hampshire red is an American breed that is derived from the popular Rhode Island red. Originally, it was bred for laying eggs, but their hearty body makes them good meat providers. Their red, single combs have five points, and their eyes are red. They make great show birds because of their colorful coat. They are versatile enough to show, to use for egg-laying, or to use for meat producing. New Hampshire reds are great for beginners because they are one of the easiest breeds to raise, due to their friendly nature and tame attitude. They are not aggressive or as flighty as other breeds and are easy to handle.

Orpington

Rooster: 10 pounds
Hen: 8 pounds
Bantam: 34 to 38 ounces
Eggs: 150 to 180 per year

This buff Orpington sits on a nest of straw.

The Orpington is a heavy, dual-purpose breed developed in England.

Currently, they are one of the most popular chicken breeds in America. The color varieties are black, blue, buff, and white. They have heavy-feathering, which makes them a good choice for harsher winter climates. Their eggs are brown. This breed is gentle and calm, which makes it a good choice for families with small children who want to be active in raising chickens. The hen can go broody if the eggs are allowed to collect in a nest.

Plymouth Rock

Rooster: 9 ½ pounds
Hen: 7 ½ pounds
Bantam: 32 to 36 ounces
Eggs: 180 to 200 per year

This dual-purpose breed was developed in the 19th century in the United States. The barred variety, which has a black and white feathering pattern, is the most popular. The white Plymouth Rock contributes to commercial broiler strains. They are excellent egg-layers of large, brown eggs. The breed is generally docile, but some birds can become aggressive.

Rhode Island red

Rooster: 8 ½ pounds
Hen: 6 ½ pounds
Bantam: 30 to 34 ounces
Eggs: up to 300 per year

This Rhode Island red shows its beautiful, dark red feathers.

The Rhode Island red is common farmyard dual-purpose breed. The breed was developed in Rhode Island and Massachusetts in the late 1800s from a number of breeds, including Malay, Shanghai, Java, and brown Leghorn. It has dark red feathers and a single-lobed comb. These distinctive characteristics helped the breed become one of the most popular breeds in the country. Rhode Island red hens lay large brown eggs, and they are a heavy meat bird. Hens can begin laying eggs when they are as young as six months old. This chicken is a good forager and is fairly docile. Consider this breed if you are a beginner because they tend to be hardy animals able to withstand a range of living conditions and diets without ceasing to lay eggs.

Wyandotte

Rooster: 8 ½ pounds
Hen: 6 ½ pounds
Bantam: 26 to 30 ounces
Eggs: 240 per year

A golden-laced Wyandotte pecking at feed.

The Wyandotte is a good dual-purpose breed. The breed was developed in the United States in the 1870s. This breed comes in eight recognized color and feathering patterns. They have a rose comb and yellow, clean legs. They are a docile, talkative breed of chicken and are a popular show breed. Wyandotte chickens are good foragers and concerned mothers.

Ornamental Breeds

Poultry enthusiasts raise ornamental breeds for use in poultry shows. These breeds are appreciated for their beauty or uncommon traits. They are not often used by commercial breeders for meat production or egg laying but can provide an interesting pet for your family. These breeds can still provide your family with eggs and meat, however – if you do not get too attached first. Raising an ornamental breed of chicken also adds to the conservation efforts of breeders.

Cochin

This blue partridge cochin is perched on a fence rail.

Rooster: 11 pounds
Hen: 8 ½ pounds
Bantam: 28 to 32 ounces
Eggs: 80 to 120 per year

This Chinese breed is a favorite for poultry shows. They have feathered feet and their feather colors can be black, white, buff, or partridge. Cochin chickens can live well in confined conditions. It is a heavy breed, with roosters weighing up to 11 pounds. The hen only lays medium, brown eggs for a short period of time but makes an excellent mother. She will even become a foster mother to chicks of other breeds.

Silkie bantam

This silkie is a free-range rooster.

Male bantam: 36 ounces
Female bantam: 32 ounces
Eggs: 50 to 120 per year

This beautiful ornamental breed originated in China in the 1200s. The Silkie comes in many colors and has feathers similar to the texture of fur, unlike the feathers of other breeds. They typically have a **top knot**, which is a poof of feathers, on the top of their head or their face can be completely covered with feathers, similar to a shaggy dog.

Silkies come in a variety of colors: red, buff, white, black, splash (a splash of another color typically highlighted on the head and back), cuckoo (barred or striped with another color such as black and white), and lavender. The Silkie has black-pigmented skin and bright blue earlobes. Their eyes are blue or black. The Silkie is also unique from other chickens in that it has five toes on each foot as opposed to the four found on most breeds.

Silkies are docile, cannot fly, and make wonderful pets for families with children. Both hens and roosters have good parenting skills; in fact, roosters often call to the chicks when food is found – in most breeds, this responsibility is solely the hen's.

Frizzles

Rooster: 11 pounds
Hen: 8 ½ pounds
Bantam: 4 to 6 pounds
Eggs: 80 to 110 per year

This frizzle cochin has a distinctive appearance.

This bird is from Southeast Asia, and is named for its feathers, which curl outward. The colors range from red, black, white, blue, silver-gray, splash (spotted), and buff, and their eyes are red. While this is a breed of chicken, the term "frizzle" is also a way to describe a bird that has fuzzy-looking feathers. Frizzles have a single comb on top of their head and clean legs that do not have any feathers. In addition to being a popular show breed, Frizzles are good egg-layers. Eggs are cream or tinted in color and of a medium size. Frizzles have a friendly demeanor and docile temperament, making them excellent additions to your family. They are easy for novice chicken owners. Their feathers do not fare well in wet weather, so they need a dry coop. They do not mind being confined, so they could be kept indoors.

Polish

Rooster: 6 pounds
Hen: 4 ½ pounds
Bantam: 2 to 3 pounds

Eggs: 125 to 150 per year

Also called a Padua or Poland, this chicken is most known for its tall, starburst-shaped crest engulfing its head. Although the origin of the breed is not well-known, statues with resemblance to the Polish chicken have been found from Roman times, and books from the 1800s discuss maintenance of the breed. They are available in several colors and markings. Solid colors are typically white, black, blue, and cuckoo, which is also referred to as barred or striped with a different color. Some birds of this breed have lace markings and are gold, silver, and chamois. The crests can sometimes cover the chicken's eyes, so they can be startled easily. It is recommended to use a special waterer, such as a nipple drinker, so the crest stays dry and clean. Polish chickens have a V- or horn-shaped comb. Care for this fowl requires more work than some other breeds because of their coat. They are partial to mites because of the thickness of their topknot. One way to prevent this is to use a spray insect-repellent, being careful not to get any in the bird's eyes. They should be kept dry because their thick feathers will hold the moisture longer. The hens are good layers and produce white eggs. They are not good sitters, though, and have been known to abandon or destroy their eggs.

Yokohama

Rooster: 6 ½ pounds
Hen: 5 ½ pounds
Bantam: 2 ½ to 3 pounds
Eggs: 40 to 60 per year

This is a striking bird with a long tail that can grow up to 2 feet in length. They are white and red-saddled or red-shouldered – meaning the feathers that cover their shoulders and upper back are a strikingly different color than their body feathers. Their skin is yellow. These are ornamental birds but are not good layers or typically used for meat because they are not fleshy birds, nor are they tender to eat. Their combs are single and thin or walnut shaped and red in color. They require a taller coop and higher perch than most breeds because of their tail. In general, the males of this breed tend to be aggressive and dominating. This breed is not recommended for a novice, but as you gain more experience they would be an excellent ornamental breed to show.

Japanese bantam

Male bantam: 26 ounces
Female bantam: 22 ounces
Eggs: 40 to 60 per year

This bird is a true bantam – meaning it does not have a large counterpart. The color varieties are white, black, black-tailed white, black-tailed buff (rusty), barred brown red (red body with brown stripes), gray, and wheaton. They have a single, red comb, red earlobes, and black eyes. Their legs are short and clean with four toes. The most notable feature of this bantam is its tail. The body is petite, but the tail is large and often reaches over the chicken's head. The males can sometimes be disqualified from competition because their tails are rye, or fall to the side rather than stand erect. The females have profuse tails but not as large as the males. The Japanese Bantam is not a proficient egg-layer, and the eggs they lay are tiny and rare. They make good pets because they are easy to tame, will not ruin your yard and landscaping, and are social birds that will interact with people.

Chapter 3

RAISING CHICKENS

Once you have decided to raise chickens and which breeds you want to start your poultry enterprise with, you will need to plan for your new arrivals. Chickens can be raised in almost any building, provided it is draft-free and predator-proof. Windows and doors to the building should be airtight and constructed of well-maintained building material. Chicks still in the down-stage — when they are covered with down and not feathers — cannot adequately insulate themselves, while fully feathered chickens can fluff their feathers, creating air pockets to protect against wind and colder weather. Predator proofing a building can be a bit trickier: Rats or weasels can wiggle through small openings, even those under 2 inches in diameter. Check the building walls and foundation for any possible openings. Seal any opening with securely nailed boards, sturdy small gauge wire, or concrete.

Predators are a big problem with all ages and varieties of chickens. Chicks need to stay inside a shelter both day and night for both temperature regulation and to keep them safe. Raccoons, cats, weasels, dogs, foxes, coyotes, skunks, and prey birds all enjoy eating chickens. Older chickens can venture outside, provided they have an enclosed shelter to escape predators. Most savvy chicken raisers will always lock their chickens in a secure shelter overnight.

Chick Care

Chicks, like any newborn animal, will need some special care to ensure they get off on the right foot and minimize losses. While all chickens should have fresh food and water, a chick is unable to regulate body heat while covered in down. Therefore, they need a constant source of external heat. Their immune systems are also less capable of fighting off disease, so pay extra attention to keeping their environment clean and reasonably sanitary. Most people find they have the most success at starting a flock of young chicks during the spring when the weather starts to warm up. This is especially recommended for northern climates. In fact, most hatcheries and feed stores will not hatch eggs or sell chicks prior to spring.

This mother hen is keeping her chicks warm.

Purchasing

You will want to start with purchasing your chicks from a reputable source whose top priority is hatching healthy chicks. Mail-order companies or feed stores are fine, provided they can tell you the source of their chicks. You will want to make sure the chicks were hatched in a reputable hatchery, not in a backyard with questionable sanitation standards. You can purchase from a breeder in your area. Just be sure to observe the health of their adult chickens to ensure they are a reliable source of baby chicks.

Hatcheries specializing in poultry are the main suppliers to feed stores or mail-order companies. Usually you can order directly from the hatchery provided you purchase a minimum number of chickens. Another added feature to investigate is whether the hatchery participates in the National Poultry Improvement Plan (NPIP) and if the breeding flock is tested yearly and certified disease free. NPIP is a voluntary program between the federal and state governments and the poultry industry to prevent the spread of poultry diseases. The hatchery will gladly tell you if they participate in this program and will be able to provide you with the necessary paperwork to assure you that their chicks are in the best possible health.

Chicks are sent when they are one day old because they still have enough nutrients from their shell to survive the journey.

Under the NPIP, the hatcheries test the blood of their breeding flock for *Salmonella enteritidis, Salmonella pullorum,* and typhoid. These diseases are passed directly from the hen to the chick. To find out the hatcheries in your state that participate in this important health concern visit: **https://npip.aphis.usda.gov/npip/openParticipantSearch.do**.

After deciding on which breed(s) of chickens you want to raise, you will next decide if you want to purchase cockerels (males), pullets (young hens under one year of age), or a straight-run. A straight-run is an order of chicks with an approximately equal amount of male and female chicks. Their gender has not been verified because the shipment is of day-old chicks. Your order will depend on how you plan on using your chickens. For example, if you plan to butcher your chickens for meat, you will want all cockerels because males will put on more meat than females. Another term for male chickens raised for meat is **broiler**. Generally, they will reach butcher weight at around 50 days of age, yielding a 4 to 5 pound carcass. This does depend upon breed and how much feed you fed them during the growing stage. Of course, you can feed them until they reach heavier weights, but 4 to 5 pounds will give you plenty of meat to enjoy.

If your plans include owning a laying flock, you will want to purchase pullets. When they start to lay eggs at around six months, they are then called hens. You do not need a rooster in order to get eggs. Hens will lay eggs without them being fertilized. The first eggs the pullet lays will be quite small; these are called pullet eggs. As the hen matures, the eggs will become larger. Either type of egg is perfectly acceptable to eat, although you may need to adjust baking recipes if your chickens are laying their first pullet eggs.

Scientists at the Roslin Institute of the University of Edinburgh recently discovered that one in every 10,000 chickens are gynandromorphous, or half female and half male. One half of the body will have rooster characteristics such as broader breast muscle, larger leg spurs, and a more pronounced comb, while the other half will have hen characteristics such as less muscling and a smaller spur and comb. The feather coloring will frequently be different on each half of the body. While this may not be more than an interesting conversation starter for the average chicken farmer, it may help researchers in poultry science to develop hens that have greater muscle mass than normal. Unlike mammals where hormones influence the development of female and male characteristics, birds do not differentiate into male and female by hormones. Instead, sex is determined by characteristics of the body cells. This finding may also help scientists determine why females, on average, live longer than males.

Once you decide the sex(es) of your chickens, it will be time to order your chickens from the hatchery. The mail-order company will tell you when your chickens will be shipped and will generally require you to order a minimum number of chicks to provide enough body heat for shipping; this is usually 25 or more chicks. Also, shipping is conducted primarily in spring and early summer to take advantage of the warmer weather. Be sure to have someone available to immediately unpack the chickens and put them in their pen when they arrive at your doorstep.

One male cock and two female hens sitting on a fence.

If you decide to purchase your chicks from a feed store, be sure to examine the condition of their pen. If the pen, water, and feed are dirty or if there are any dead chicks in the pen, do not purchase your chickens from that store. The chicks' health might already be compromised from a poor start in life. It is also important to check the vent (anus) of the chicks. If there is any fecal build-up, the chicks might be harboring a disease. The eyes and nose should be clear and free of any discharge. If a baby chick remains huddled when the rest of the chicks are roused, it is probably sick and should not be purchased. In fact, you may want to pass on purchasing any chicks from this particular pen, as the other chicks may also come down with whatever disease the sick chick is suffering from.

Equipment

You can purchase heat lamps and bulbs at farm supply stores or hardware stores. Heat lamps are round metal lamps and are fairly inexpensive — some cost less than $5. The bulbs are generally 125- to 250-watt bulbs that produce an ample amount of heat and light for keeping the chicks warm. Other types of heat lamps are also available; Nasco (**www.enasco.com**) carries a wide variety of heat lamps and bulbs for poultry farmers. Another option is to purchase a gas, oil, or electricity-fired hover brooder. A hover brooder is a heat lamp with a canopy to keep warm air close to the ground. The cost of these types of hover brooders is not effective if you have fewer than 100 chicks. This brooder also does not provide light to see the chicks, but it is effective at keeping the young birds warm and toasty. Regardless of heat source, you need to provide at least 7 square inches of space beneath the brooder or heat lamp per chick.

A heat lamp is important for chicks that are not being kept with a mother hen.

Chicken-waterers should hold 1 gallon of water or less and should be slightly elevated after the first few days to prevent buildup of shavings and manure from the chicks' natural tendency to scratch and kick bedding into the water or feed trough. You can purchase a smaller waterer for chicks to use while they are young and unable to withstand the cold. Likewise, feeders should be raised slightly after the first few days. Supply one water provider per 25 chicks and 1 inch of feeder space per chick. Waterers or feeders can be raised off the ground using bricks or pieces of wood. A simple wire stand can be made using four pieces of lumber nailed together in the shape of a square. Cover this frame with a small square of sturdy wire, nailed securely to the boards with U-

These hens are eating from a feeding trough.

shaped staples. This will keep the water and feed off the ground and minimize contamination.

For the first few weeks, the chickens can be raised in a large stock-tank or similar container. A stock-tank is a large container generally used to hold the feed of large animals, like cows. The main priority is that the container be cleaned, bleached, and dried before the chicks arrive. A mild bleach or vinegar solution (1 teaspoon per gallon of water) can be used to disinfect the pen, waterers, and feeders. Let the equipment dry in the sun; this ensures all disease-causing germs are killed.

Preparing your brooding pen

Spread a 2- to 4-inch layer of bedding on the floor or in the bottom of your container, and then hang your heat lamp. Make sure the heat lamp or brooder will not be in contact with any combustible materials and that it is securely fastened. You will need to adjust the heat lamp or brooder based upon chick behavior so you will want the lamp to be easily moved. Start out with hanging the lamp so the bottom is 12 inches from the bedding.

Fill the waterers with fresh water. You can add a pinch of vitamin and electrolyte powder to the water for the first week to give your chicks a good start. This medicine can be purchased from any general farm store or your local veterinarian's office. It provides the chicks with vitamins A, D, E, B, and C, along with electrolytes. This will give the chicks an added boost after the physical stress from being shipped to your farm. Every day mix up a fresh batch of water with this supplement. While directions for use will vary, typically you will add 1 teaspoon per each gallon of water. Always read the package directions prior to use. Durvet Inc. makes a good vitamin and electrolyte mix (**www.durvet.com**).

Fill the feeder or feed trays with chick starter, which can be purchased at farm stores or a local **grain elevator**, where grain farmers sell and store their grain. Another first day tactic is to spread the feed on newspaper or cardboard and place this on the floor. The chicks will then be able to readily find and eat the feed immediately, which is an important factor, as they will be hungry. After they have eaten, remove the newspaper and cardboard and observe them to make sure they are eating out of the feeders.

When your chicks arrive, take each one out of the box and dip its beak in the water. This will ensure that the chicks get a drink of water and that they will know where the water source is located. Then, release the chicks under the heat lamp. After all chicks have been released, observe them at least five times over the course of the day to make sure they are all able to eat and drink. If they huddle under the lamp, the lamp should be lowered to provide more heat. If they are scattered far away from the lamp, they are probably too warm, and the lamp should be raised a few inches. This is a good general rule for most young animals.

A covering of chicken wire or other woven wire over the container – even if your building is predator proof – will give you extra assurance that nothing will sneak into the pen and kill your investment. If you have a shallow container, it will also help contain the chicks as they get older and are testing their abilities to jump and flutter.

Designing Your Chicken Coop

After your young chicks are fully feathered they will need to be moved to an adult chicken house. The basic requirements for a good chicken coop will protect them from the weather and predators, as well as allowing them to room to carry out their essential functions: growing, feeding, and egg-laying. The coop can be almost any structure you can think of, but there are some requirements for a good chicken coop.

Coops can be any size, shape, color, or composed of almost any building material. The size of your flock will determine how large of a coop you will need and what quantity of building material you will need to build your coop. Materials used to build a coop can be purchased at lumberyards, hardware stores, through online sources, at flea markets, or at Habitat for Humanity stores. To find a local Habitat for Humanity store, you can go to its national website, **www.habitat.org**. If you plan on using recycled lumber, do not use wood that has been painted or chemically treated with preservatives. Chickens will peck at everything as they explore their environment and may inadvertently ingest paint chips or pieces of treated wood. The basic requirements for a versatile chicken coop include:

- Solid construction to protect against predators and weather
- Bedding and a good floor
- Nest boxes
- Roosts or perches
- Feeders and waterers
- Ample lighting
- Adequate ventilation
- Insulation, especially in northern climates.

This chicken coop is set on cinder blocks and constructed of wood.

Solid construction

The average adult chicken will need 3 to 4 feet of space per chicken, and bantams require 2 square feet per chicken. If you plan to breed your chickens, they will need more space. Building a pen of an adequate size will prevent overcrowding, which can lead to disease, aggression, and cannibalism. You will want to find an appropriate space on your property prior to building or purchasing your coop. A good spot will have the following:

- Adherence to zoning laws, such as keeping your coop away from your property line
- Enough room to construct the coop
- Good drainage so standing water does not collect under the coop
- Sunlight

Your coop should have a solid roof made of metal panels, roofing shingles, or composite panels. A tarp will not provide sufficient protection and will soon wear out due to weather exposure. The sides can be made of a welded wire material, but if you live in a cold-weather state, solid walls with a few windows that can be opened will provide the best protection against weather and predators. Chicken wire is not adequate for your coop as predators can tear through the wire in their quest for chickens and eggs.

Chickens are hardy in winter, but need a solid coop for protection.

Bedding

The chicken coop floor can be composed of concrete or wood. Some chicken owners will cover the floor with linoleum or vinyl to help with cleaning and disinfection. Regardless of the flooring type, bedding should be used. Shavings are best as they absorb moisture well and give the coop a pleasant smell when freshly applied, but hay and straw can also be used. Bedding should be monitored and changed when it becomes visibly soiled. Under no condition should manure be allowed to build up because the fumes given off can adversely affect your chicken's health.

Little Creek Farm in Ocala, FL uses straw bedding.

One method of bedding is to deeply bed the coop with 4 to 8 inches of bedding material. To maintain the bedding you will need to rake it frequently to evenly distribute the material and to help aerate the bedding. This bedding, if properly maintained and if you prevent overcrowding by providing sufficient room, can be kept in place for a month before it needs to be changed. If the coop starts to smell, you will need to remove all the bedding, clean and disinfect the coop, and new bedding material should be placed in the coop. The used bedding makes excellent compost for gardening.

If you want to clean the coop more frequently you can use 2 to 3 inches of bedding and remove the bedding as soon as it becomes soiled. You will not need to completely clean and disinfect the coop after each removal, but it is recommended to do so at least twice a year to control disease.

Nest boxes

Eggs freshly laid in a nesting box.

You will need to provide your hens with a **nesting box** that has clean bedding material. Hens need a secluded place to lay their eggs and will lay them in undesirable or hidden spots if not provided with a comfortable nesting box. You can use individual boxes or a community box. You can put four hens in one nesting box, or 1 square foot of space per hen is needed for a community box. Individual nesting boxes should be at least 12 inches wide by 12 inches high. As a general rule, be sure your hen can stand in the box comfortably. Community boxes can be any size but need at least two 9-inch by 12-inch openings. The openings should be provided with a flap of material covering 2/3 of the opening to provide seclusion. The nest boxes should be securely fastened to the side of the coop, two feet above the floor. They must be lower than the roosts because a bird will roost in a nesting box if it is the highest place in the pen. You can build your own boxes using untreated plywood or boards. These can also be purchased through feed stores or online sources such as **www.enasco.com**. The nest boxes should be cleaned as soon as the bedding material becomes soiled. Eggs can quickly become dirty, and the hens may lose interest in laying eggs in a dirty nest box.

Lighting

Lighting is important in the chicken coop. It allows you to be able to observe the conditions inside the coop easily, and it is also essential for egg production because hens need a certain amount of light each day in order to stimulate egg laying. Under natural lighting conditions, egg-laying is maximized during the spring and summer months when there is more daylight. It tends to drop off dramatically in the winter when daylight is short. Lighting needs will vary depending on growth stage. Here is a great guideline on how much light chickens should receive:

- Up to 7 days old: 24 hours a day
- 1 to 6 weeks old: 8 to 12 hours a day
- 6 to 19 weeks old: 12 hours a day
- 20 weeks and up: 12 to 16 hours a day

Ventilation and insulation

Good ventilation is absolutely necessary inside the chicken coop because it allows fresh air and oxygen into the coop, as well as releases heat and harmful gases, such as ammonia and carbon dioxide. Windows can allow fresh air inside the coop. An alternative is to design the coop with a 2-inch ventilation opening right under the roof. This opening should be enclosed with sturdy welded wire to prevent the entry of small flying birds, bats, or predators. Insulation is also important to keep the chickens warm in the winter. This will help retain body heat from the chickens or from any supplementary heat you provide. Along with insulation you should put a vapor barrier to prevent condensation and ice build-up along the walls. Cover any insulation with plywood or thick plastic to prevent the birds from destroying the insulation with their pecking.

Roosts or perches

Roosts are important for laying flocks. Chickens have a natural instinct to perch off the ground to avoid predators. This is especially important at night when they are sleeping. Do not use roosts or perches for birds used for broilers as they may cause breast blisters to form on the chest. A breast blister is a swelling of the tissue in the sternum that occurs in

A propped ladder makes a great place for your birds to roost.

poultry breeds, and a blister will often lead to a downgrade in the meat. Allow 6 to 8 inches of roost space per bird for the smaller breeds. Larger breeds should have 8 to 10 inches of roost space. Roosts will also allow manure to accumulate in one area, making it easier to clean up, and a well-designed coop will have a manure pit beneath the roosts. This is an area covered by a simple screen made of wood or plastic and covered with welded wire.

These chickens are ready to roost.

Roosts should be placed away from the nest boxes, feed, and waterer equipment to prevent manure contamination. The roost should be constructed of 2-inch by 2-inch. wood with rounded edges. Roosts can be arranged horizontally or in a ladder-style arrangement. An old wooden ladder can also be propped and secured to the wall, or old wooden handles from brooms or gardening tools also make ideal roosts. If you use more than one roost, they should be placed 12 inches apart from each other. Be sure the roosts in the coop are higher than the nesting boxes or else the birds will use the nesting boxes for roosting.

These chickens are in a pen with a feeder and waterer.

Feeders and waterers

The same types of feeders and waterers provided for young chicks can be used inside the chicken coop. Feed should only be fed inside the coop to keep other animals from eating the feed and contaminating the feed with their droppings. Waterers can be placed inside or outside. During colder weather, water will need to be closely monitored for freezing. In cold climates a heated waterer will prevent this problem from occurring. Hang or place the feeders so that the area where the feed accumulates is level with the chickens' backs. This will prevent excess feed waste, as chickens love to scratch at the feed. It will also help prevent bedding accumulation as chickens also enjoy scratching through the bedding.

Where to get your coop

After you evaluate all the coop requirements you will be ready to build. You can design your own coop based upon the necessary requirements. There are many available designs online to help you with building. Here is a good list of online coop designs. Some are free while others will need to be purchased.

- **Green Roof Chicken Coop (www.greenroofchicken-coop.com)**: This website offers plans for sale, pre-cut coop kits, and assembled coops that you can order.

- **BuildEazy (www.buildeazy.com)**: This website offers a variety of build-at-home projects. You can access free, step-by-step instructions, which will include a list of materials and photos of what your project should look like.

- **Braingarage (www.braingarage.com)**: This website provides free plans for a chicken coop, along with a variety of building plans for other projects. This website offers tips, designs, and some plans you can download free of charge. People share their experiences about building chicken coops and how well their design works for them.

Of course, you can also purchase premade coops if you are not a good carpenter. Here are a few websites that sell chicken coops:

- **Egganic Industries (www.henspa.com)**: They offer both portable and stationary chicken coops in a wide variety of styles.

- **Amish Goods (www.myamishgoods.com)**: Many different styles of coops and options are available from this manufacturer.
- **Chicken Coop Source (www.chickencoopsource.com)**: The site has a variety of coops as well as other products you might need for your new chickens.

A way to give your chickens exercise, to control insects and bugs, and to allow your chickens access to fresh grass is through the use of chicken runs and portable chicken coops. Chicken runs are enclosed areas attached to the chicken coop where chickens are allowed to exercise. At a minimum the run should have 10 square feet of space per chicken, but more room is optimal. Runs are usually constructed of wood posts to which chicken wire is securely attached. While the top does not necessarily need to be closed, covering it with mesh will keep out birds of prey like owls and hawks. The chickens should be securely locked inside the coop at night because predators are apt to attempt to break into the run. For small flocks, a tractor coop may be an ideal way to combine a coop and a run. These bottomless coops are mobile and can be moved around the barnyard or an orchard to give your chickens access to fresh grass and to control insects. Remember not to put the chickens in an area that has been freshly fertilized or that has had chemicals applied.

Farmer cooperatives are places where farmers can purchase items they need in bulk to save money. You do not need to become a member in a co-op to shop at one, but if you are, you enjoy added advantages because shopping here gives back locally to your community. Here are some ways you can locate a farmer co-op:

- Your local Yellow Pages.
- **Co-operative Feed Dealers (www.co-opfeed.com)**: Established in 1935, Co-operative Feed Dealers, Inc. is a distrib-

utor for independent farm, garden, and pet stores in the northeastern United States.

- **Local Harvest (www.localharvest.org)**: This website allows you to type in your ZIP code or city and state and will help locate farmers' markets, farms, places to show for sustainable and organic products and food, and give information on community-supported agriculture.

- Your local county extension service.

Supply stores are usually staffed by well-educated employees and can greatly assist you in making your selection. If they do not have a product you are looking for, they can often help you find it at another location or special order it for you. Here are some ways to locate a supply store near you:

- Your local Yellow Pages or classified ads in local newspapers.

- **Tractor Supply Company (www.tractorsupply.com)**: This website offers general farming and agriculture supplies, and it mostly sells products, but does offer information on farming.

- **Horizon Structures (www.horizonstructures.com)**: This website offers products for farming and agriculture, and it has a variety of coops to choose from.

Farmers' markets are great places to find bargains and unusual items. If you are purchasing your eggs or birds here, ask several questions, especially about the age of the pullet, because a hen's laying ability is directly related to her age. Buying a coop here gives you an opportunity to discuss raising chickens with someone who has experience. The downside is that there are no refunds or returns on products you buy here.

A Sample Chicken Coop Plan

Chicken Coop Plans

Bill of Material

Item	Qty	Description	Item	Qty	Description
1	6	2x4x3'-0"	16	1	1/2" plywood x 1'-2" x 1'-8"
2	6	2x4x4'-0"	17	4	1/2" plywood x 1'-2" x 1'-2"
3	4	2x4x8'-0"	18	1	1/2" plywood x 1'-0" x 2'-0"
4	2	2x4x2'-9"	19	2	1/2" plywood x 2'-7" x 8'-0"
5	1	2x4x3'-9"	20	4	2" hinges
6	1 box	3" deck screws	21	1	1 latch
7	1	1/2" plywood x 3'-0" x 4'-0"	22	6	1x1x8"
8	1 box	1 5/8" deck screws	23	1	2x4x2'-0"
9	1	1/2" plywood x 2'-11" x 4'-0"	24	2	2x4x4"
10	1	1/2" plywood x 2'-11" x 4'-0"	25	1 bdl	Shingles
11	1	1/2" plywood x 1'-7" x 3'-0"	26	1 box	1" Roofing Nails
12	1	1/2" plywood x 1'-7" x 3'-0"	27	1 roll	Wire Mesh
13	1	1/2" plywood x 1'-2" x 3'-1"	28	1 box	1" heavy duty staples
14	1	1/2" plywood x 1'-0" x 3'-1"	29	1 gallon	primer
15	1	1/2" plywood x 1'-4" x 3'-3"	30	1 gallon	paint

Step 1: Assemble Base Frame:

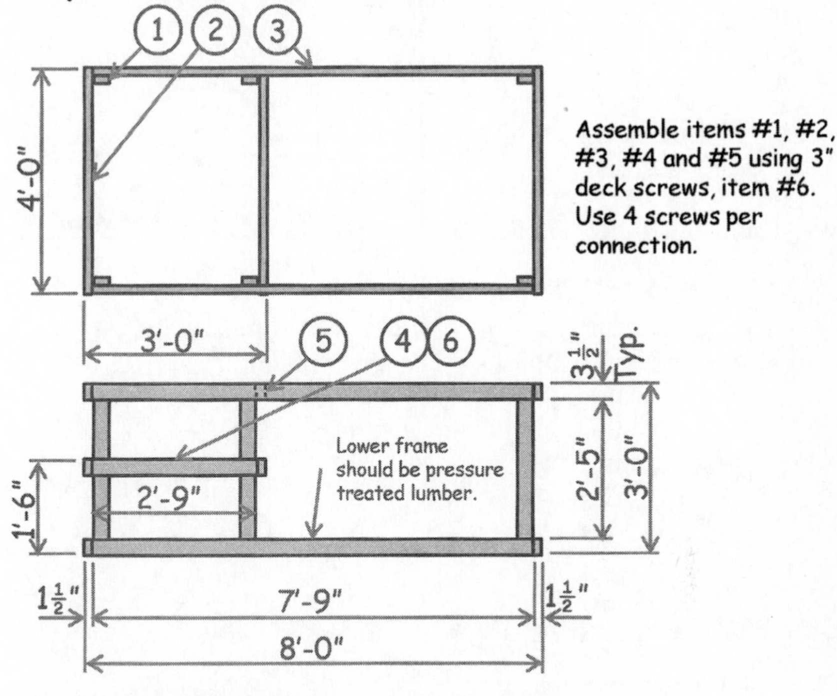

Assemble items #1, #2, #3, #4 and #5 using 3" deck screws, item #6. Use 4 screws per connection.

Step 2: Cut Out and Install Coop Floor:

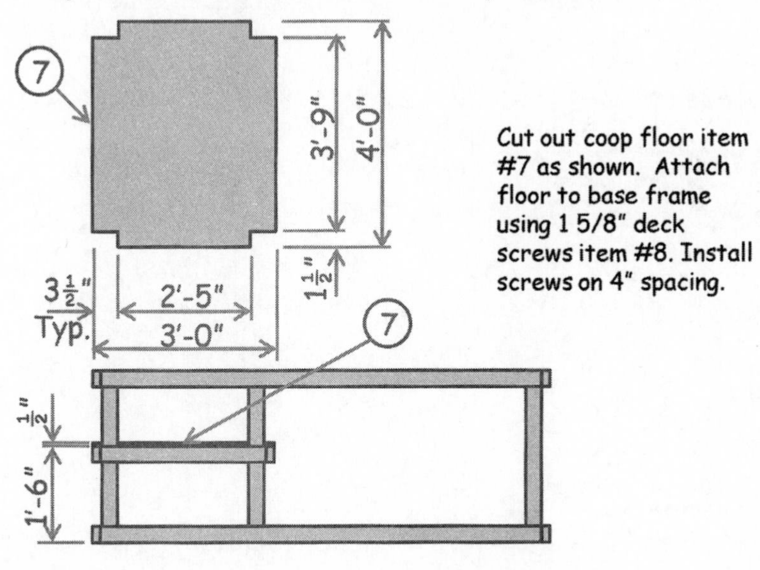

Cut out coop floor item #7 as shown. Attach floor to base frame using 1 5/8" deck screws item #8. Install screws on 4" spacing.

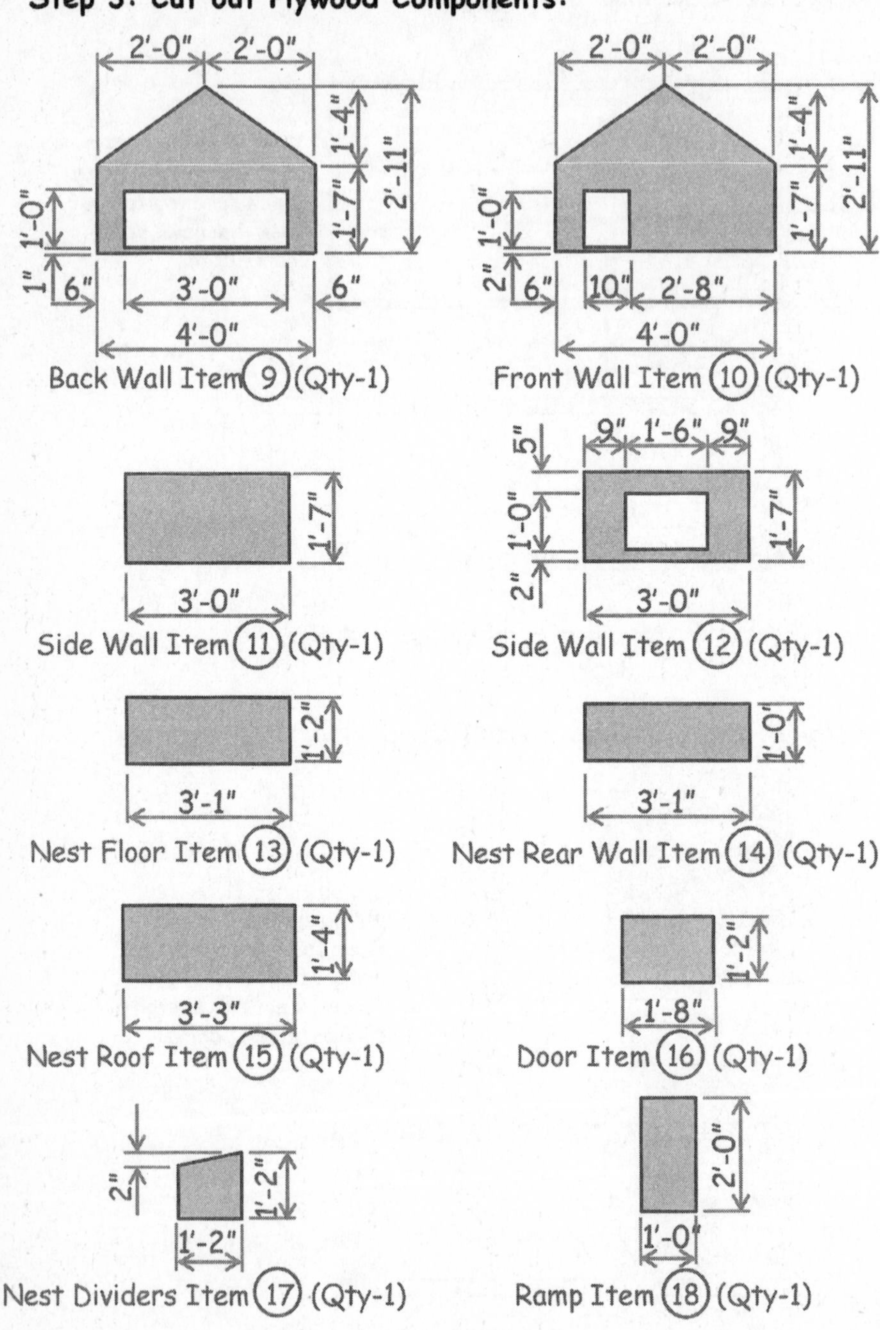
Step 3: Cut out Plywood Components:
2'-0"
2'-0"
1'-4"
1'-7"
2'-11"
1'-0"
1"
6"
3'-0"
6"
4'-0"
Back Wall Item 9 (Qty-1)
2'-0"
2'-0"
1'-4"
1'-7"
2'-11"
1'-0"
2"
6"
10"
2'-8"
4'-0"
Front Wall Item 10 (Qty-1)
1'-7"
3'-0"
Side Wall Item 11 (Qty-1)
5"
9"
1'-6"
9"
1'-0"
1'-7"
2"
3'-0"
Side Wall Item 12 (Qty-1)
1'-2"
3'-1"
Nest Floor Item 13 (Qty-1)
1'-0"
3'-1"
Nest Rear Wall Item 14 (Qty-1)
1'-4"
3'-3"
Nest Roof Item 15 (Qty-1)
1'-2"
1'-8"
Door Item 16 (Qty-1)
2"
1'-2"
1'-2"
Nest Dividers Item 17 (Qty-1)
2'-0"
1'-0"
Ramp Item 18 (Qty-1)

Step 4: Assemble Nest Box:

Assemble nest box items #13, #14, & #17 using 1 5/8" deck screws item #8. Install screws on 4" spacing. Space item #17 dividers equally.

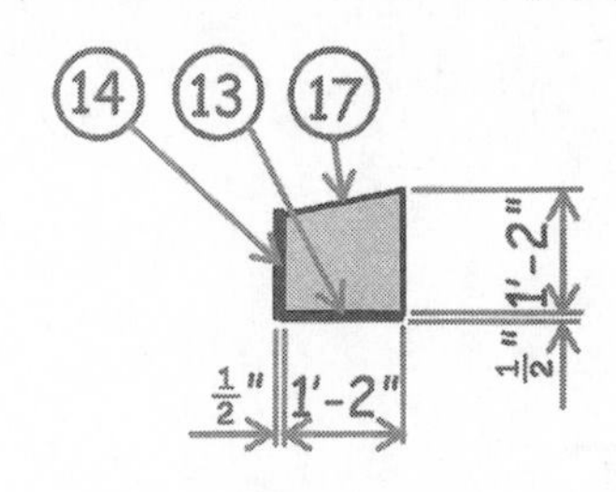

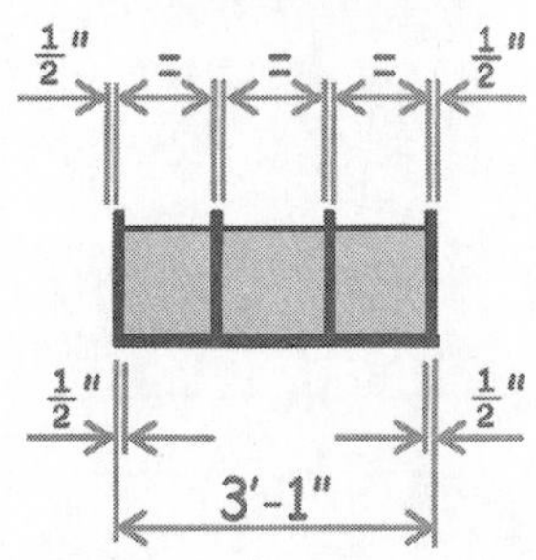

Step 5: Attach Walls and Nest to Base Frame:

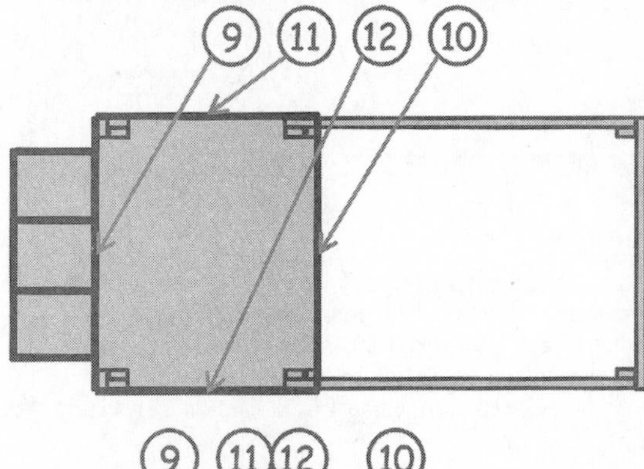

Attach coop walls to base frame using 1 5/8" deck screws item #8. Attach nest box items #13, #14, & #17 to rear wall using 1 5/8" deck screws item #8. Install screws on 4" spacing. NOTE: Nest floor should be 2" lower than coop floor.

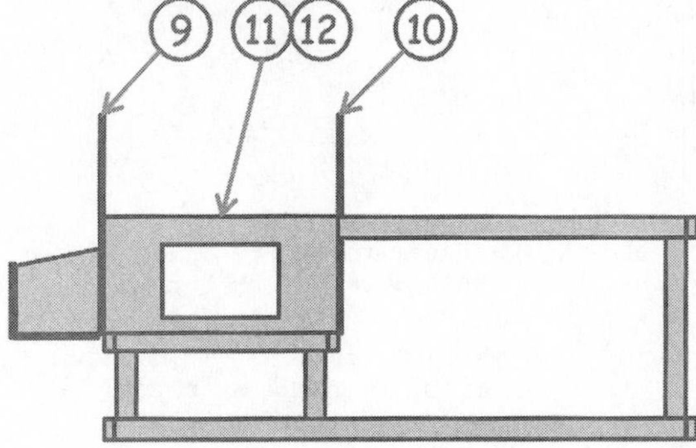

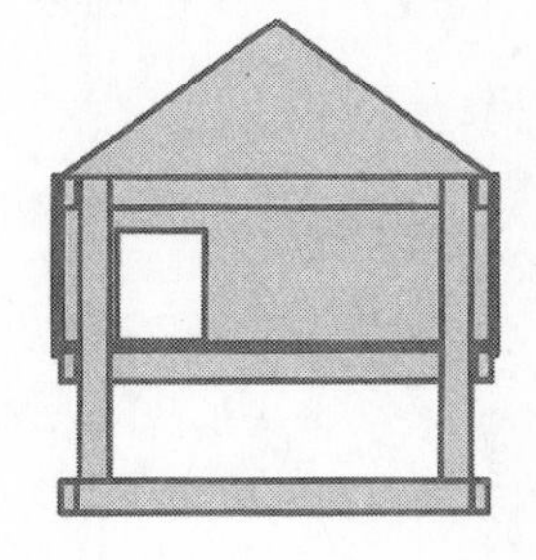

Step 6: Install Roof, Door, Nest Lid, and Ramp:

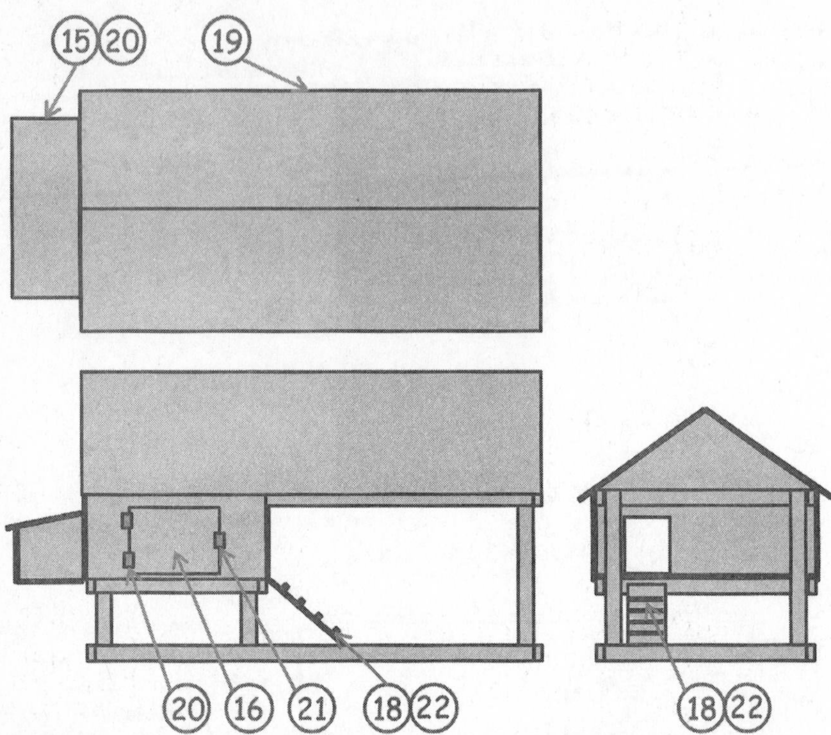

-Attach roof item #19 to walls and base frame using 1 5/8" deck screws item #8.
-Attach nest box lid item #15 to rear wall using 2 hinges item#20.
-Install access door item #16 complete with latch item #21 and hinges item #20.
-Attach ramp item #18 to base frame using 1 5/8" deck screws item #8.
-Attach cleats item #22 to ramp, equally spaced, using 1 5/8" deck screws item #8.

Step 7: Assemble Perch:

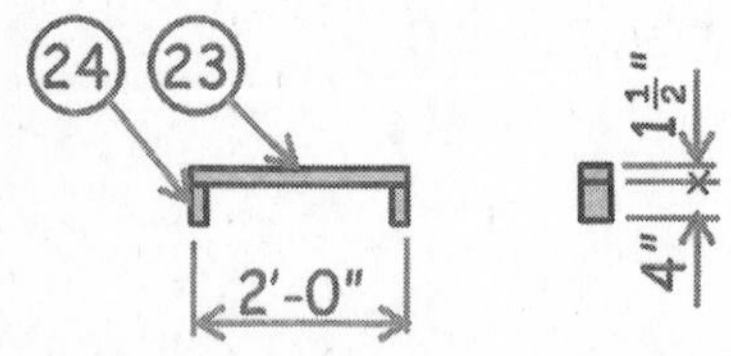

Assemble perch items #23 & #24 using 3" deck screws item #6. Use 4 screws per connection. Attach perch to coop floor using 3" deck screws item #6. Locate 6" from front wall of coop and 3" from side wall, parallel to front wall.

Step 8: Install Roofing Shingles, Wire Mesh, and Paint to suit:

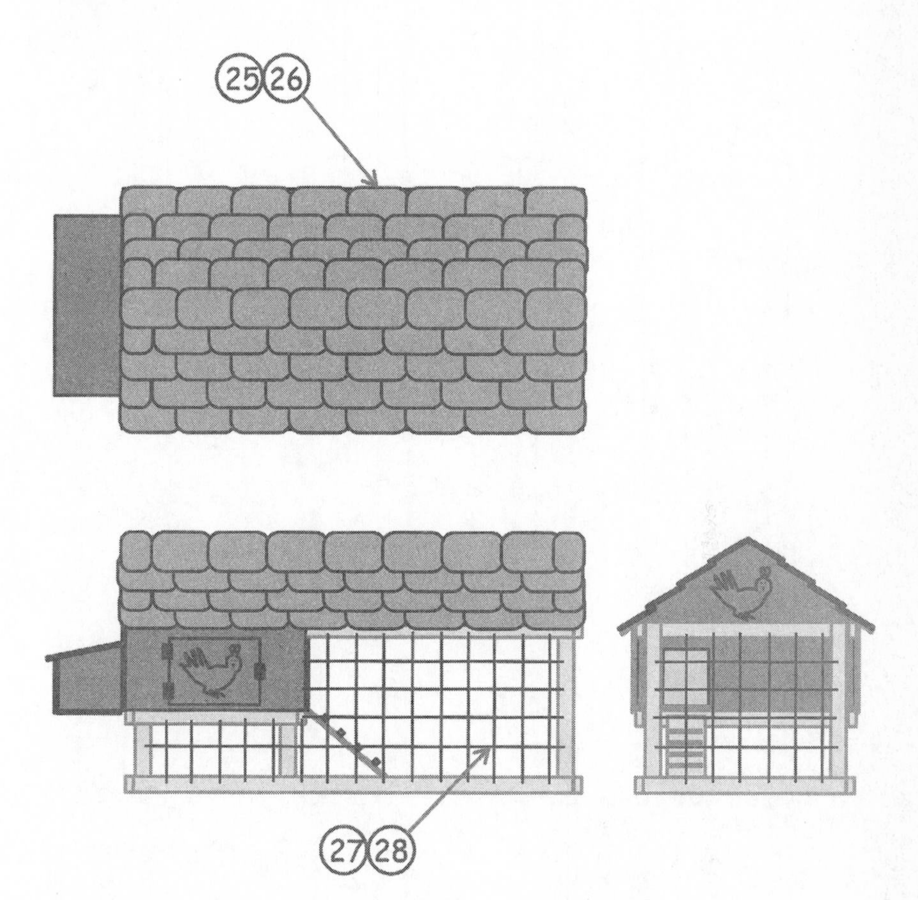

-Install roofing shingles item #25 to plywood roof using roofing nails item #26. Follow manufacturers instructions.
-Attach wire mesh to exterior of base frame using 1" heavy duty staples.
-Prime and paint to suit.

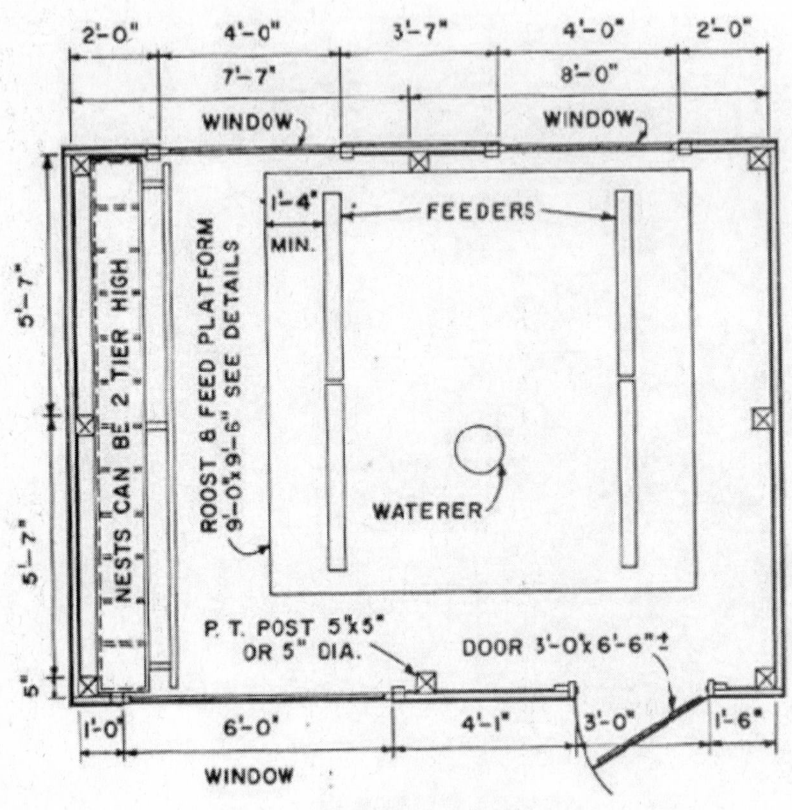
2'-0"
4'-0"
3'-7"
4'-0"
2'-0"
7'-7"
8'-0"
WINDOW
WINDOW
FEEDERS
1'-4" MIN.
ROOST & FEED PLATFORM 9'-0"x9'-6" SEE DETAILS
NESTS CAN BE 2 TIER HIGH
5'-7"
5'-7"
5"
WATERER
P. T. POST 5"x5" OR 5" DIA.
DOOR 3'-0"x6'-6"±
1'-0"
6'-0"
4'-1"
3'-0"
1'-6"
WINDOW

FLOOR PLAN

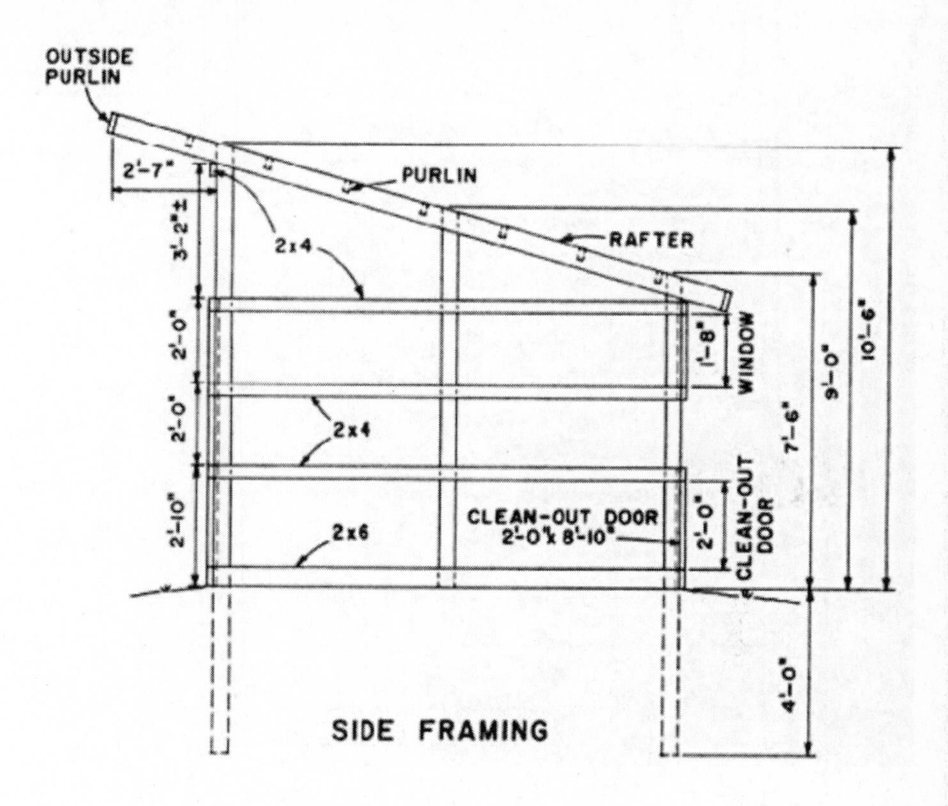
OUTSIDE PURLIN
2'-7"
PURLIN
2x4
RAFTER
3'-2"±
2'-0"
2'-0"
2'-10"
2x4
2x6
1'-8"
WINDOW
CLEAN-OUT DOOR
2'-0" x 8'-10"
2'-0"
CLEAN-OUT DOOR
7'-6"
9'-0"
10'-6"
4'-0"
SIDE FRAMING

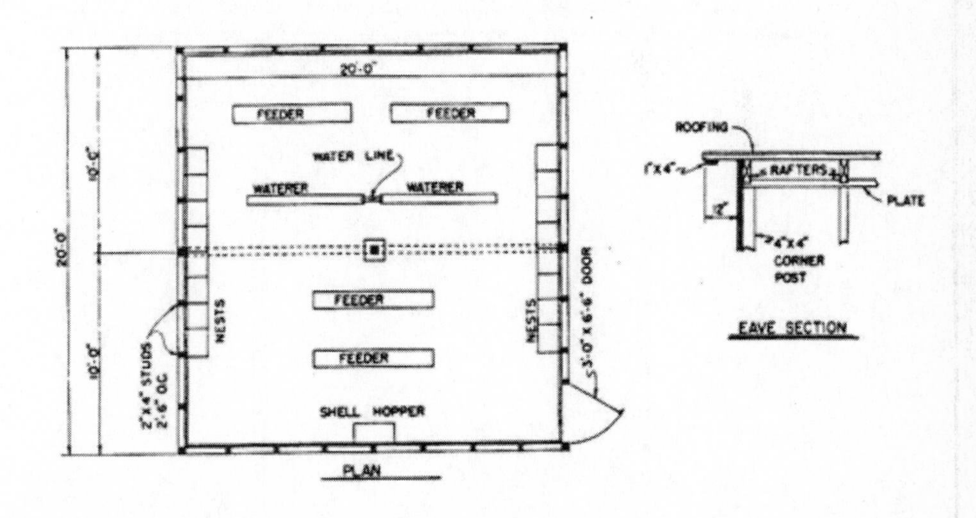
20'-0"
FEEDER
FEEDER
WATER LINE
WATERER
WATERER
10'-0"
20'-0"
10'-0"
NESTS
NESTS
FEEDER
FEEDER
2"x4" STUDS 2'-6" O.C.
3'-0" x 6'-6" DOOR
SHELL HOPPER
PLAN
ROOFING
1"x4"
RAFTERS
PLATE
12"
4"x4" CORNER POST
EAVE SECTION

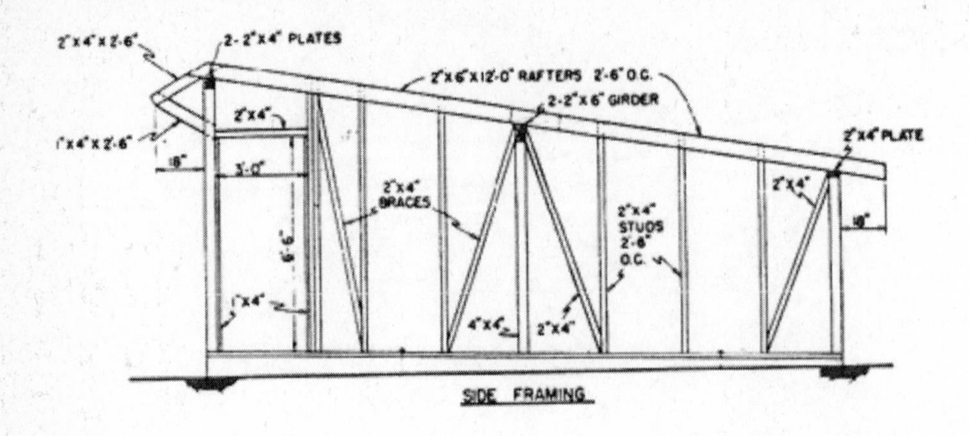

2"x4"x2'-6"
2-2"x4" PLATES
2"x6"x12'-0" RAFTERS 2'-6" O.C.
2-2"x6" GIRDER
2"x4"
1"x4"x2'-6"
2"x4" PLATE
18"
3'-0"
2"x4" BRACES
2"x4"
2"x4" STUDS 2'-6" O.C.
6'-6"
18"
1"x4"
4"x4"
2"x4"
SIDE FRAMING

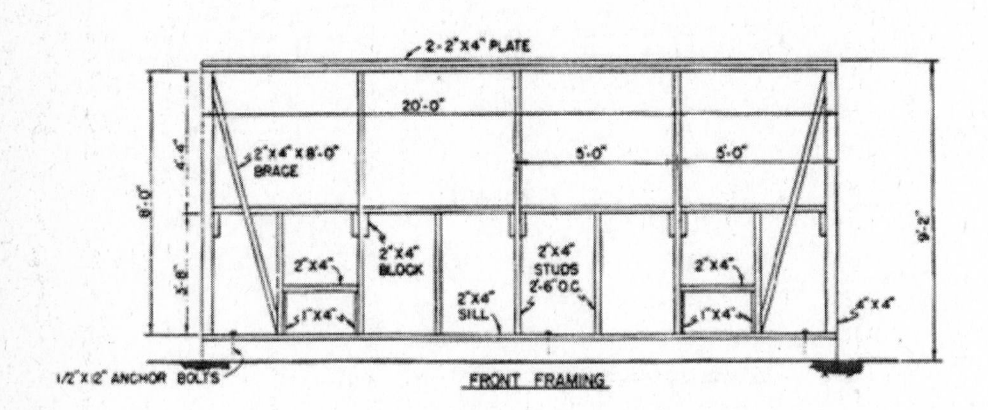

2-2"x4" PLATE
20'-0"
2"x4"x8'-0" BRACE
5'-0"
5'-0"
4'-4"
8'-0"
3'-8"
2"x4"
2"x4" BLOCK
2"x4" STUDS 2'-6" O.C.
2"x4"
2"x4" SILL
1"x4"
1"x4"
4"x4"
9'-2"
1/2"x12" ANCHOR BOLTS
FRONT FRAMING

Plan for an 8' x 8' Layer House - 15 to 20 Hens

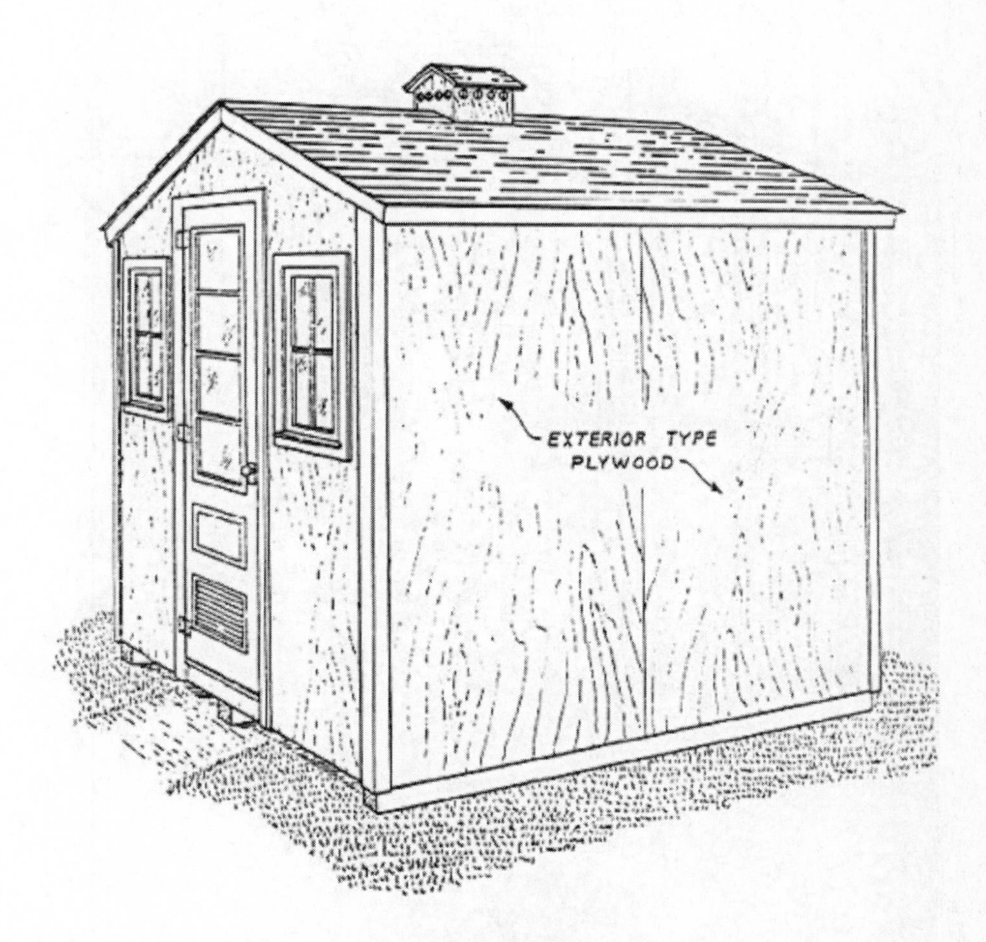

PERSPECTIVE

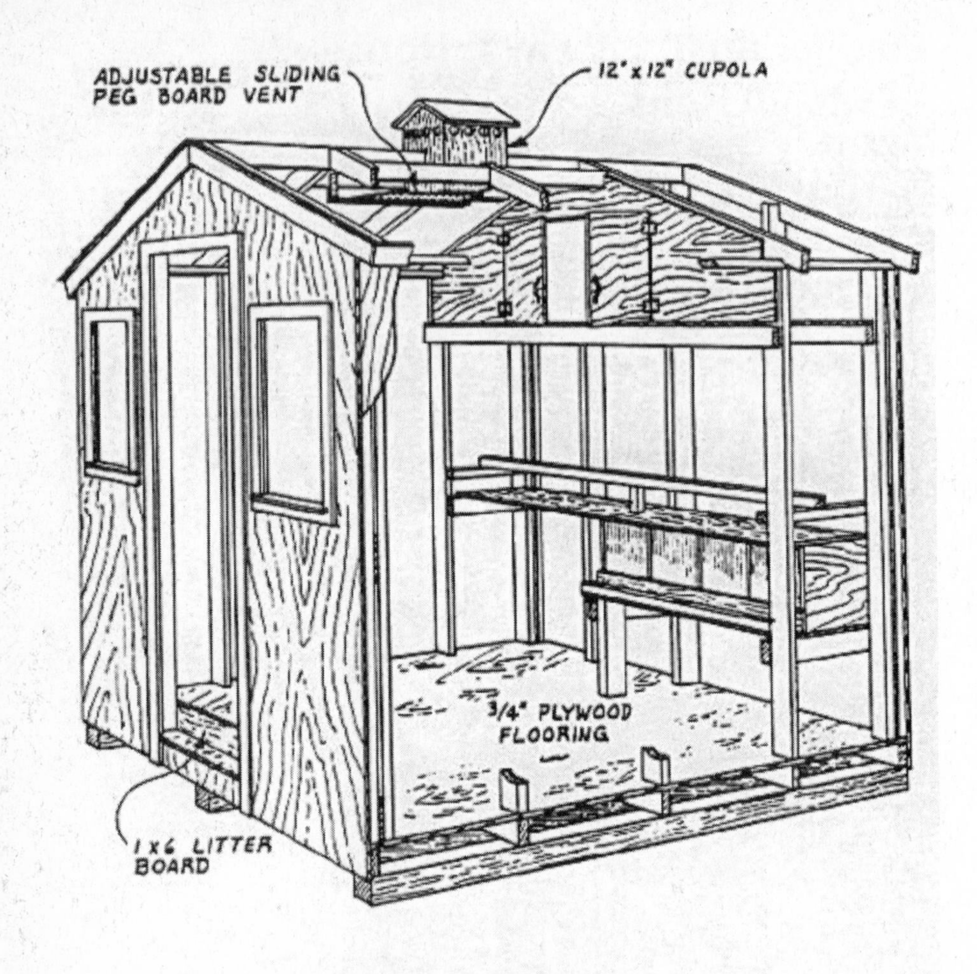
ADJUSTABLE SLIDING PEG BOARD VENT
12" x 12" CUPOLA
3/4" PLYWOOD FLOORING
1 x 6 LITTER BOARD
CUTAWAY VIEW

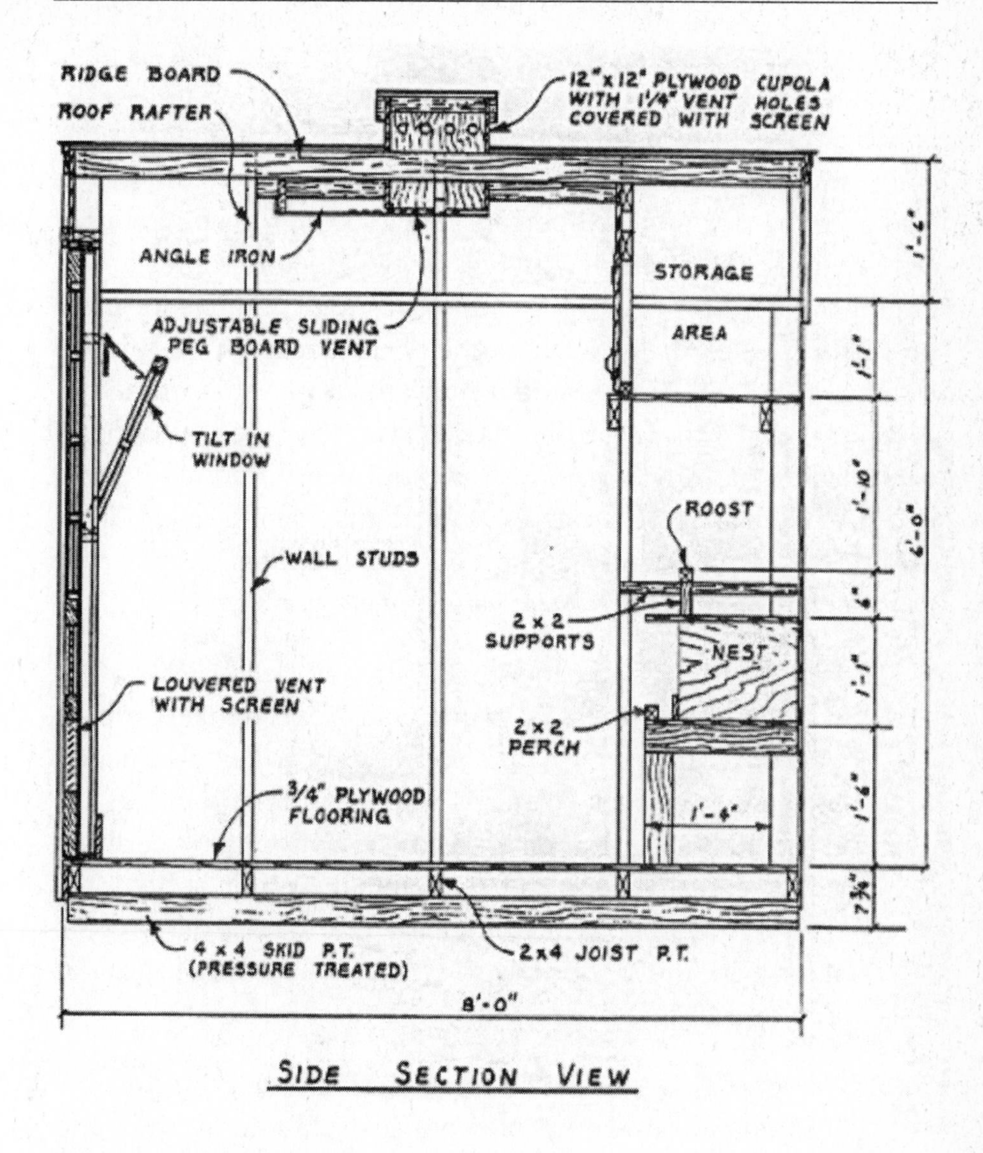

Reviewed by Audrey McElroy, associate professor, Animal and Poultry Sciences

Nutritional Requirements

Having the best feed for your chickens will allow them to develop properly and prevent many diseases and conditions. It also accounts for 60 to 70 percent of the cost of keeping chickens. There are different feeding strategies depending on if you are raising hens for egg production or raising cockerels for meat production. Regardless of what type of chickens you are raising, though, there are some essential nutrients every chicken needs. A lack of any of these nutrients will adversely affect all aspects of the chicken: growth rate, reproduction, egg production, and egg size. Here are the six classes of nutrients that are essential for life, growth, reproduction, and production. *Note that every breed of poultry in this book requires these same six nutrients.*

1. **Water:** A lack of a consistent source of clean, fresh water will slow the growth of young chicks and reduce production of meat birds. It will also lead to low egg production in the laying flock. Water is important for proper digestion and body metabolism. There is a strong link between water consumption and feed intake. Water softens the feed in the crop (feed storage sac in the body) to prepare it for grinding in the gizzard where coarse grain is ground down prior to entering the stomach.

2. **Protein:** This is essential for growth and egg production. Proteins are made up of more than 23 amino acids and chickens require certain levels of each amino acid. Protein is usually provided in the diet from animal sources (meat meal, fishmeal, or milk waste) because this form is more effective than proteins from grains. Soybeans also provide valuable proteins. Too much protein can negatively affect chickens as well, so it needs to be carefully monitored.

3. **Carbohydrates:** These provide energy for chickens and make up the largest part of the poultry diet. They are the

starchy or sugary materials found in grains. Too many carbohydrates will make a chicken fat.

4. **Fats:** These also provide energy and are good to feed chickens during cold weather when they use more energy to keep warm. They are also essential to the absorption of vitamins A, D3, E, and K. In addition, fats are needed for reproduction and egg formation. Too many fats will cause digestive and nervous system problems.

5. **Minerals:** Chickens use minerals like calcium, phosphorus, and salt in small quantities to fire enzymatic reactions throughout the body, to form strong bones, and to form the eggshell. Other minerals needed in the diet include manganese, iron, copper, zinc, and iodine. Grains are low in minerals so poultry feeds need to be supplemented with minerals to make sure the chickens get all they need.

6. **Vitamins:** Vitamins are important to grow young poultry, and chickens need 13 vitamins in their diet. These vitamins include vitamins A, D3, E, K, B12, and the B vitamins: riboflavin, thiamin, nicotinic acid, folic acid, biotin, pantothenic acid, choline, and pyridoxine. Vitamin A is essential for the health and functioning of the skin and digestive tract. Vitamin D3 is needed for bone formation. The B vitamins are needed for metabolism throughout the body.

You have an array of feeds to choose from, or you may decide to make your own. Remember not to purchase too much feed at once because you should not store it over long periods of time; it can attract rodents, and if it gets moist, it can become moldy.

A group of free-range chickens feeding.

You can find chicken feed at local feed mills, supply stores, co-ops, and even some grocery stores, depending on where you live. Chickens are natural scavengers, and they will peck and hunt daily for food. But even free-range chickens will need you to supplement their diet with feed unless you have an abundance of land for them to forage. It is estimated that an acre of land can sustain 50 to 400 chickens, but that is probably not accurate in the winter months when greenery is hard to find.

You may be tempted to buy a cheaper feed for your chickens, but be aware that the product is mostly bulked up with fillers such as wheat-milling byproducts. This will fatten your birds up but has little or no nutritional value. It is like junk food for poultry and will affect its flavor. A fat chicken will lay fewer eggs than a healthy chicken.

This rooster is feeding in a free-range environment.

Chicken feed comes in three forms:

- **Mash** is a mealy or powdered form of chicken feed, usually made of grains.
- **Pellet** is a harder form of chicken food.
- **Crumble** is a mixture of mash and pellet.

Chickens tend to waste more mash because it spills to the ground and dissolves into the dirt. Chicken scratch was a popular feed before people realized how important nutrients were to the quality of the chicken and the eggs it produced. The scratch was made up of whole

These chickens are sneaking some feed.

grains and cracked corn. If you use chicken scratch, it should not be their only feed because it does not provide enough nutrients, such as oyster shell, to harden the chickens' eggs.

Chicken feed also comes in different mixtures. Look around the feed store and you will see feed for different chicken ages and purposes: baby chicks, pullets, layers, broilers, and crushed oyster shell, among others. Oyster shell helps the development of the eggshell, and if a chicken does not have enough in its diet, the eggs it lays will be brittle or soft. Commercial feed mixtures will be balanced so you do not have to add anything to them.

Making your own feed may save you some money, but it will require more time. You will need:

- Split peas – these have a high protein content
- Lentils – a plant from the legume family producing flattened seeds used as food and a good source of protein
- Oat groats – minimally processed oats
- Hulled barley – these help provide intestinal protection
- Sunflower seeds – benefit the heart
- Sesame seeds – rich in vitamin B and E
- Flax seeds – this provides omega-3
- Winter wheat – higher gluten protein than most wheats
- Whole corn – this provides energy for flock
- Soft white wheat – this is lower in protein and higher in carbohydrates compared to hard wheat
- Quinoa – good source of dietary fiber
- Kelp granules – gives the birds potassium, iron, and fiber

- Oyster shell – provides calcium and helps build hard egg shells
- Granite grit – this helps to aid digestion in birds
- Millet – most nutritious of the grains, providing proteins, carbohydrates, as well as phosphorous and other nutrients
- Kamut – this is known as a high-energy wheat

Mix equal parts of the ingredients, except corn (use two parts), soft white wheat (three parts), and hard red winter wheat (three parts). Mix and store in an airtight bin.

Even though chickens need feed, they love bugs, worms, vegetables, and table scraps. Watermelon rinds are also a delightful treat. Chickens self-regulate their food, meaning they will not overeat, so you can let them have access to their feed all day.

Some farmers used cracked corn for feed.

Because chickens do not have teeth, they eat grit, which is stored in the crop area of their stomachs. Grit is little stones that break down the food so the chicken can digest it. You can find it at the same places you purchase chicken feed. You can add it to food or give it separately. Grit is not the same as oyster shells; oyster shells dissolve in the chicken's stomach, and grit does not.

Starter rations

The young chick needs a diet rich in protein and vitamins, along with a balanced mineral mix. Two pounds of chick starter will feed one laying breed chick for about six weeks. Broiler chickens will consume about 8 pounds of feed during this time period. After six weeks, the chick should be switched to a broiler or grower ration, depending on if the chicks are intended to be meat-type or laying birds. Chickens intended for laying will need the grower ration. You do not want the layer to grow as rapidly as a broiler because you want to delay sexual maturity until the bird is mature, which is at around six months.

A meal or flake ration should be purchased because a finely ground mix will tend to paste in the chick's mouth. The medicated chick starters will have a coccidiostat, which is a drug added to help control coccidiosis. Coccidiosis is an infection often in the intestinal tract. You may not realize a chick is ill until you see bloody droppings. The disease can be prevented with cleanliness and use of medicated chick-starters. Non-medicated chick starter will work as well, as long as you are diligent about daily cleaning of waterers and feeders and do not overstock the pen. There is no medication that can take the place of cleanliness and good husbandry, which is the practice of raising stock — including poultry breeds.

Chicks have special dietary needs.

Broiler ration

This feed should be available at all times for the chickens to maximize growth rates. The ideal feed will have 20 percent protein along with the other essential nutrients. Many feeds will contain a coccidiostat. Feeding this medication should be discontinued ten days prior to butchering to allow time for the broiler's body to rid the meat of this medicine.

Grower ration

This ration is fed to those birds intended for breeding or egg-laying. This diet will have moderate levels of energy and 14 to 16 percent protein. Feed this diet to your breeders until the chickens are about 5 months of age when they should be switched to the laying ration.

Laying ration

Laying rations need to provide all the necessary nutrients as listed above. A balanced diet of 16 to 17 percent protein should be provided to the layers. You can also provide whole grains such as oats, wheat, or corn in a separate feeder for the hens. They should also be provided grit, a hard, coarse material that helps the hens grind the feed properly in the gizzard.

The layers can also be fed kitchen or garden scraps but do not feed them strong smelling scraps like onions as these might give the eggs a bad flavor. Many people will also provide the hens with a separate feeder with calcium carbonate – either through purchasing crushed oyster shells or using ground up eggshells – to provide extra calcium for proper eggshell formation. Be sure your breeders do not get too much calcium because excess calcium can damage the kidneys. A diet needs to be 1 percent calcium for a chicken to be healthy.

Breeder diet

The breeding diet is basically the same as the laying diet, but it will need to have more riboflavin and vitamin D than the laying ration. These are essential for fertility and hatchability of the eggs. The breeder diet should be fed two months prior to saving the first eggs for hatching.

Clipping Wings

There is something so restricting when you imagine a bird with clipped wings – it seems unnatural to alter what comes natural to birds by preventing them from taking flight. The truth is, though, when owning chickens, some breeds are able to fly, and if your fence or surroundings are not built to keep them secure, clipping their wings may be one solution. Because the neighbor's dog or a two-lane highway may be within their reach, clipping their wings can keep them safe and keep them alive. In some cases, clipping a bird's wings helps to keep the bird free-range because you do not have to keep it enclosed most of the time.

If this is your first time clipping a bird's wing, you may want to enlist the help of an experienced friend. If you do not know of anyone who has chickens and has done this before, find a friend or family member that can help you hold the bird during this process.

Clipping your chicken's wing should not hurt the bird. It is like clipping your dog's nails. Most often, the animal is afraid of the process. For birds, the vibration from the actual clipping of their stiff feathers scares them. There should be no blood or very minimal blood. If you clip your chicken's wing, and the bird starts to bleed, use your first aid kit to stop the bleeding. If the bird is bleeding heavily, call or visit your vet immediately to treat the bird.

You can clip a chicken's wings when its adult feathers are grown in. Even if a chicken has its adult feathers, it is advisable to wait until flight becomes a problem. Do not cut pinfeathers, which are the tips of new feathers on a bird's body. Clipping one wing is enough to prevent the chicken from taking flight; it throws the bird off balance and still allows it to fly, just not very high. Some chicken owners prefer to clip both wings to keep the bird balanced. The wings will grow back, just like people cut their hair and it grows. There are tutorials of how to clip wings posted on YouTube (**www.youtube.com**) by chicken owners. Go to the website and type in the search "How to Clip a Chicken's Wings."

To get started, you will need:

- An assistant
- Very, very sharp scissors
- Old towel
- Rubber gloves to protect your hands
- A first aid kit (just in case of injury)
- Treats for when the task is complete.

1. First, decide who will clip the wings and who will hold the bird.

2. Gather your chicken. Be calm. If you cannot get hold of the chicken, it is not advised to chase it around. This stresses the chicken. Try again at another time or try to gather another bird.

3. Hold the chicken by the legs. Support its body by keeping your hand underneath its body. Your palm should be open and flat. Leave either the left or right wing free.

4. Talk to it in soothing tones.

5. Spread the wing. Display it in its entirety.

6. The first ten feathers from the outside of the wing are flight feathers; these flight feathers are usually longer than the rest and often a different color. Cut just beyond the edge of the next layer up. This is approximately 3 to 5 inches.

7. Quickly clip the bird's wings. Use very sharp scissors. Dull blades may hurt the bird.

8. Give your bird a treat.

Chickens molt annually, so you will need to clip your chickens' wings each year with its new growth of feathers.

Chapter 4

BEHAVIOR AND PRODUCTION MANAGEMENT

Behavior

While chickens, like all birds, are visually-oriented, they can hear quite well. They do not have an external ear, like mammals, but they do have a round opening on the side of their head that is hidden by tiny feathers. Chickens communicate using sounds and gestures. If you take time to observe a flock of hens, you will find that they spend a lot of time "talking" to each other. Hens have a range of vocal calls of more than 30 different sounds. They will alert other chickens if they find food or

This colorful chicken in an Indonesian market is showing its personality.

bugs with one call, while a mother hen will speak to her chicks with a completely different sound. Right before laying an egg, a hen will give an almost distressed call, while after laying the egg she will give a proud cackle to announce the new arrival.

Pecking and cannibalism

Pecking and cannibalism are major problems in chickens. Chickens have a social need for a pecking order formed within the flock. The rest of the flock will defer to the top bird, giving it access to food, water, and space first. Usually the top bird in a flock is a rooster or an older hen. A flock of chickens will establish a social hierarchy and the hierarchy will coordinate flock activities such as drinking, eating, dust bathing, and roosting. Every chicken knows its rank in the order, and the rank seldom changes unless an illness or death occurs.

A rooster will be at the top of flock hierarchy.

Cannibalism can occur in chickens for a variety of reasons, but it often occurs due to their environmental factors or hierarchy problems. It is important to make sure your chickens have the right amount of lighting, a proper diet, supplemental nutrients, disease control, and enough space to live. This will prevent cannibalism in the flock. Chickens peck one another when they are establishing their hierarchy, but the pecking can lead to cannibalism if the environment is not healthy.

Commercial growers will trim newborn chicks' beaks to cut down on pecking and cannibalism. Beak trimming will not allow your chickens to forage effectively if your plans include letting your chickens outside to eat grass, bugs, and insects. As long as you are not restrictively confining your hens to tiny, commercial-type pens, pecking and cannibalism should not be a problem and beak trimming will not be necessary.

Chickens have good eyesight and can see colored objects in bright sunlight. This will lead them to use their sensitive beaks to explore new food. If their beaks become soiled or if they eat mushy food, they will wipe their beaks off on the ground. Their beaks can be compared to our hands because they use the beaks to explore and carry things. An intact beak allows the chicken to fully explore their world.

Aggression

If you have a particularly aggressive bird that attacks the other birds by feather-pulling or picking at other chickens' rear-ends, it can be placed in a separate pen but within sight of the rest of the flock for a few days to help it calm down. It can then be reintroduced to the flock, but be sure to monitor its behavior. Placing extra feed and water sources in the pen will also cut down on competition for these resources.

Many times aggression will rear its head when new chickens are introduced into an established flock. If you must bring in new chickens, a good method is to remove the old flock members from the pen and place them into a new, temporary holding pen for one week. Place the new arrivals in the old pen. After a week, put the old chickens back with the new. Another method is to physically divide the pen in half and putting the new chickens in one half along with feed and water. In a week or two, the chickens will have acclimated to the new arrivals and the barrier can be removed.

If, despite all your attempts to tame aggression, you find one bird has been de-feathered by the other birds, there are bad tasting and smelling ointments that can be placed on her feathers to deter pecking by the other chickens. The chicken may need to also be removed from the flock temporarily to heal. Reintroduce her to the flock via the methods given above. Finally, there are plastic devices called bit rings that can be placed on the beak of particularly aggressive birds, which will allow them to eat and drink but not peck the other chickens. You may find pushing down the rear-end of the aggressive chicken and holding the chicken to the floor for a minute in the presence of the flock may help deter bullying behavior.

This rooster is standing among its flock of hens in a pasture.

Taming a mean rooster

Mean roosters have turned many people off from raising chickens. With their well-developed spurs, rooters can inflict serious wounds when they see you as a threat to their flock dominance. Each time you enter the coop, you will have to remind your rooster that you are the head of the flock.

An alpha rooster usually dominates the pecking order of a flock by getting first dibs at food, water, and hens. The hens also have an alpha hen that is the boss. Unlike the roosters, hen social order is harder to see, but it is there. The alpha rooster asserts his dominance constantly. If another chicken attempts to eat first or another rooster tries to mate a hen, the alpha male will rush over and administer a good peck or even pummel the out-of-bounds interloper with beak, claws, and wings.

A rooster in the hen house.

Your job is to make sure you are not the victim of such an attack. Starting when the chickens are young, take a few minutes each day to observe their behavior. When the chicks start to fight (or spar), break them apart by gently pushing them back with your fingers. Do the same thing for each fight you see. This will let the flock know that you are in charge. As they get older, you will still need to reinforce your dominance. It is fine to pick up a rooster and pet him, but a rooster should not be eager to approach you. If he does, he will think that he is the top chicken and not you. It will not hurt to take an occasional swat at him if he seems too comfortable.

This rooster, named Crackle, is being held, showing submission.

Do not let the rooster eat before the hens. Doing so will give him the cue that he is dominant in the flock but establishes your dominance over him. In essence, you are top rooster and are allowing your hens to eat before him, the less dominant rooster. If he tries to come to the feeder first, push him away so the hens eat first. After the hens are eating, it is fine if the rooster begins to eat. If a rooster tries to breed a hen in your presence, push him off and chase him away. It is the same concept as eating; the dominant rooster (you) has first rights to everything. He has to wait until you leave to be able to breed the hens.

Roosters can be aggressive and need special management.

The main thing when it comes to roosters is that you have to be on guard at all times. If these techniques do not work, or if you are unable to be consistent in asserting your dominance, it might be safer for the rooster to be destined for the dinner table.

Managing Your Birds

Breeding

Before you embark on breeding your chickens, take some time to decide what your intentions are. Do you want to have replacement layers and some broilers to feed up? Are you interested in breeding purebred chickens or one of the uncommon breeds? Do you hope to sell baby chicks? Breeding chickens and incubating eggs will take some planning on your part to ensure your chicks are healthy and are able to survive.

Most hatcheries have eggs for sale in the spring and early summer to coincide with favorable weather for raising chicks. This is especially important in the northern latitudes, where harsh winter weather makes keeping a brooder area at the proper temperature in unheated and poorly insulated buildings nearly impossible. You should aim for a spring hatch if you live in a colder area. You will be hatching out both males and females. If you breed a laying-type chicken, the males will not make the best meat-birds. They can be raised for meat, although most likely at a financial loss. Meat-type female birds are not necessarily the best layers, although they can be butchered for meat. Hybrid birds, as opposed to purebred strains, will not necessarily look like their parents. So choose your breed based on your desired outcome.

Hens like this brown one need a special diet for egg production.

Cockerels reach breeding age at 17 to 19 weeks of age, and one rooster can service between six and 12 hens. Hens reach breeding age at 18 to 20 weeks. The hens and rooster should be placed in an indoor pen with nest boxes (one box per four hens) so you can collect the eggs twice daily. The collected eggs can be stored in egg cartons at 55 to 60 degrees Fahrenheit for up to seven days before they are incubated. Eggs should be cleaned of manure and dirt before storage or placing in the incubator, and then they should be candled five days after being placed in the incubator. A broody hen can also sit on a nest of eggs if you do not have an incubator. She should be left alone to care for the eggs, but provide her with feed and water at all times.

Laying flock management

You will need to properly manage your laying flock in order to get a consistent, quality, and clean supply of eggs. As discussed previously, laying hens will need to be fed a steady diet of laying-hen feed, which has more calcium than the other types of chicken feed. In addition, crushed oyster shells (which come in convenient 40 to 50 pound sacks) should be placed in a separate feeder and not mixed in with the feed. Keep the feeders and waterers full of feed, oyster shells, and water at all times for your hens.

As points of reference, the top of the egg is the pointy, thinner portion, and the bottom of the egg is the wider, rounder part. The following is the anatomy of an egg, starting from the outside and going in:

- The entire egg is covered with a hard coating called the eggshell. It is made of calcium carbonate and has approximately 17,000 pores to allow moisture and air to circulate. A light coat or membrane around the egg keeps out bacteria.

- The outer shell and inner shell membranes are next inside the shell. They surround the albumen, or the egg white. They also help keep bacteria out.

- The **chalazae** are located at the top and bottom of the egg. They are long, twisted cords that attach to the shell and hold the yolk in place. A fresher egg will show more prominent chalazae.

- The **exterior albumen** is the narrow fluid layer next to the shell membrane.

- The **middle albumen** is the thick, white layer of dense matter known as the egg white.

- The **vitelline membrane** is the clear coating enclosing the egg yolk. If you hear the term *mottled* in reference to the egg, it means the membrane is covered with blotches or pale spots.

- The **Nucleus of Pander** is a plug of whitish yolk located inside the yolk that is purely for nutritional value.

- The **germinal disk**, or **blastoderm**, is a small circular spot on the surface of the yolk. This is the point where the sperm enters the egg. It will gradually send out blood vessels into the yolk once it is fertilized for nutrition as it develops.

- The **yellow yolk** is at the center of the egg and can range in color from yellow to orange, depending on the hen. The yolk makes up about 33 percent of the egg and is comprised of valuable nutrients, such as protein, calcium, iron, and vitamin B12. The older the egg gets, the larger the yolk becomes because it absorbs water and nutrients from the albumen.

- The **air cell** is located at the bottom part of the egg. When the egg is first laid, it is warm. When it cools, it contracts. The inner shell membrane pulls away from the outer shell membrane and forms an air pocket.

Hens can start laying eggs as early as 4 months old, so you can expect to have farm-fresh eggs relatively soon if you started your brood from chicks. The color of the hen's first egg will be the same color as all of the eggs she will lay throughout her life. For an easy reference, here are breeds and the color of the eggs they lay:

White eggs:

- Leghorn
- Houdan
- Polish
- Sultan
- Cornish (although some of their eggs are a creamy, white color)

Brown eggs:

- Rhode Island Red
- Wyandotte
- New Hampshire
- Orpington
- Welsummer (these are a chocolate brown egg)

Blue, green, green-blue, pinkish-brown and cream-colored:

- Ameraucana produces more colorful eggs than any other breed.
- Silkies mostly produce tinted or cream-colored eggs.
- Plymouth produce light pink to medium brown eggs.

Egg Oddities

Most eggs have only one yolk, but occasionally you will find a double yolk or even a triple yolk egg. This usually happens when the chicken is new to laying eggs and her cycle is adjusting. When ovulation starts too rapidly, an additional yolk is produced but has nowhere to go. It connects to the next yolk and ends up inside one egg. Double yolks and triple yolks are safe to eat.

Occasionally, an egg is passed through a chicken that does not have any yolk at all. These eggs are referred to as "wind" eggs or "dwarf" eggs. This also occurs more frequently in pullets rather than older hens. As the hen's cycle is just beginning, it may take a few times to get in sync with the bird's body.

Even more rare than finding extra yolks or no yolks is an egg within an egg. Sometimes you will crack an egg open only to find another egg inside. The cause is not quite clear, but it is possible the egg developed and then while it was in the duct, it either reversed direction or got stuck when the next cycle began. It is rare, but it happens.

Once in a while, an egg is dropped from a hen and does not have an outside shell, but just the yolk and white part of the egg covered in a membrane sac. This is an accident of a hen's reproductive system. You do not need to be concerned about the chicken's health if these oddities occur. If they happen regularly, take your bird to the vet or investigate its eating habits. The bird is most likely lacking nutrients.

Nesting

You will want to make sure you can find the eggs once your hens start laying. Having nest boxes in your coop is an easy way to do this. You can do this by fencing in the flock so they cannot wander off into the woods or other protected area to start a nest. Some breeds of hens are more broody than others, meaning they will lay an egg daily in a hidden nest and try to sit on them to hatch out chicks. They will even do this if there is no rooster to fertilize the eggs. A

This crele bantam hen is outside foraging.

nest box can be that sheltered, shady place for broody hens to lay their eggs. This need not be fancy or elaborate, but it does need to be maintained and cleaned frequently. Eggs can quickly become soiled by feces or mud, which can also stain the shell, leading to unappealing eggs.

There should be one nest box for every four hens in your flock. This will help to keep the eggs clean and to minimize egg breakage. Depending on the style you purchase, the boxes should have a deep layer of litter to cushion the eggs and to absorb any waste material. Nest boxes should be placed inside a building. Rats, skunks, raccoons, and snakes will eat eggs, so the building should be predator proof, especially at night. Hens will learn to come back to the pen to lay eggs, so they can be let out of the pen during the day to forage and exercise.

Eggs should be picked two times each day. Most chickens will lay their eggs in the morning. If the temperature is below freezing, check for eggs frequently as they can freeze. By picking eggs frequently, you will also help to minimize breakage. Hens will frequently peck at broken eggs and sometimes can develop an appetite for eggs, breaking intact eggs just to eat them. Once you have collected the eggs, they will need to be processed. The eggs should be washed with lukewarm water as soon as possible after they have been picked. If the water is colder than the eggs, dirt and bacteria can be drawn into the egg through pores in the shell. Using warmer water will make the contents inside the egg swell and push the dirt away from the pores. A mild detergent such as a liquid dishwashing detergent can also be used to wash eggs. Use running water to clean the eggs and do not allow them to sit in water to minimize any bacterial cross contamination between the eggs.

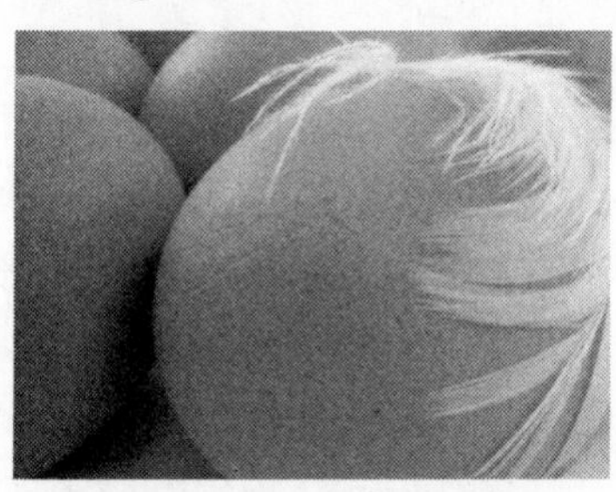

Eggs are an excellent source of protein.

Then dry and cool the eggs after washing. To properly store, they should be placed in egg cartons with the pointy end facing down, and the carton should be placed inside a refrigerator. Storing with the pointy end facing down gives the yolk a chance to settle properly. Cleaned and stored eggs can remain safe to eat for four weeks or longer.

CASE STUDY: PREVENTION OF EGG EATING

Phillip J. Clauer,
Poultry Extension Specialist,
Animal and Poultry Sciences

Egg eating by hens is a habit formed over time which is extremely difficult, if not impossible, to break. It is important you plan and manage your facilities so that the hen never gets the first taste of a broken egg.

Prevention management practices include:

1. **Reducing Traffic in the Nesting Area.** Egg breakage is a major reason why hens start eating eggs. Excessive traffic in the nesting area increases the chance of egg breakage. Some precautions which can be taken include:

   a) Provide one 12-inch by 12-inch nest for every four to five hens in your flock. Never have less than six nesting boxes. Always locate the nests at least 2 feet off the ground and at least 4 feet away from the roosts.

   b) Keep 2 inches of clean, dry nesting material in the nests at all times. Many eggs are cracked due to a lack of protective padding in nesting boxes.

   c) Remove all broody hens from the nesting area. Broody hens reduce nesting space and cause more traffic in the remaining nests.

2. **Nutrition.** To keep the egg shells strong, feed a complete ration and supplement with oyster shells. The oyster shells serve as a calcium supplement to keep the shells strong.

   Never feed the hens used egg shells without smashing them to very fine particles. If the hen can associate the shell to the egg, the hens are encouraged to pick at the fresh eggs in the coop.

3. **Keep Stress Minimized**

   a) Don't use bright lights in your coops, especially near the nesting area. Bright light increases nervousness and picking habits.

   b) Do not scare the hens out of the nesting boxes. The sudden movement can break eggs in the box and can give the hens a taste of egg and promote egg eating.

4. **Egg Eating Can Be From Outside**. Egg eating can be done by predators such as snakes, skunks, rats, weasels and other predators. If your hens are eating eggs, the hens will usually have dried yolk on their beaks and sides of their heads. Egg eating hens also can be seen scouting the nests for freshly laid eggs to consume.

   If you do catch an egg eater, cull her from the flock at once. Egg eating is a bad habit that will multiply the longer you let it continue. If one hen starts eating eggs, other hens will soon follow.

   Prevention is the only proven treatment. Collect eggs often and collect eggs early in the day. Most hens will lay before 10:00 am each morning. The longer the eggs are in the barn, the better the chance they will be broken or eaten.

Reviewed by Audrey McElroy, associate professor, Animal and Poultry Sciences

Molting your hens

Molting is a process through which hens lose their feathers. It happens naturally usually during the late fall or early winter, although the time of year can vary according to when the birds were hatched and depending upon breed. This is an important process in laying hens and breeding chickens because it will extend their egg-laying life. It usually happens after the hen has been laying eggs for nine to 12 months. During the molt, which lasts about eight weeks, the reproductive tract regresses, and it will stop laying eggs. After this time the feathers will regrow, and the hen will resume laying eggs; however, egg production will not be as high as it was during the first year of life. Molting your birds can decrease the initial cost of replacing old hens with young chicks.

A forced molt can be done if your flock is falling off in egg production. You should weigh a few birds (five is a good number) and find their average weight. Reduce light in the hen house to eight hours of light. Remove the feed but provide water. A week later, weigh your birds again. They should have lost 30 percent of their body weight to undergo a full recovery of the reproductive tract. After the weight loss, you can then provide a food ration at 50 percent of normal consumption. Two weeks later give the hens at least 12 hours of light and resume feeding them a laying ration at normal levels. Forced molting is controversial because of the feed restriction and forced weight loss, so it is not a common practice among small flock owners.

Your bird's beautiful feathers will grow back after a molt.

Candling and incubating eggs

Eggs are candled in order to see if they are fertilized or not. Despite daily gathering of eggs, occasionally you may miss one for a few days and an embryo may develop if you have a rooster running with your hens. If you plan to sell your eggs, you should candle them to make sure your customers do not crack open an egg with a chick growing inside. Eggs are also candled if you are incubating fertilized eggs to determine if the chick is growing or if it has died inside the shell.

An actual candle is not used to candle eggs. To candle the eggs, you will need a darkened room and a bright, direct light source. You can use a bright flashlight or a lamp with a bright bulb. You can also purchase a commercial candler through poultry supply companies. Hold the larger end of the egg up to the light and slowly turn it. You should be able to see the air sac, the yolk, and the pores through the eggshell. Dark or brown eggs will make it a little harder to see through the shell. In a fertilized egg, there will be a spot, a thin red ring or blood vessels, around the yolk. If the egg has been incubated about a week, you will be able to see the embryo's eye, the shadow of its body, and it may even move. If you have a dead embryo, you will see a blood ring around the yolk, or possibly a dark spot.

The website The Easy Poultry Chicken and Supply (**www.shilala.homestead.com**) offers a homemade candling system:

Materials:

- 60-watt sealed-beam floodlight bulb
- Ceramic light base
- Old lamp cord
- 4-inch-by-4-inch utility box
- Clamp connector for nonmetallic cable
- Piece of scrap wood for a mounting base
- Cardboard box with a small hole cut in it
- One roll of black electric tape

Directions:

1. Drill or cut a hole in the side of the utility box and attach the clamp connector on the outside.

2. Take the lamp cord with the bare ends ¾ of an inch and pass through the connector, leaving 6 inches of cord inside of the box.

3. Screw down the 4 by 4 box to your piece of wood that you are using as the base and tighten the connector screws.

4. Attach the lamp cord to the light base. One wire goes on each screw, it does not matter which wire to which screw.

5. Attach your lamp base using the screws that came with it.

6. Put the light bulb in. Take the cardboard box and cut a small hole in it and start candling. The hole only needs to be large enough to see the egg. Do not leave the candler unattended because the bulb gets hot and could start a fire.

A fun family project is to incubate eggs, either purchased or laid by the family hen, until they hatch. To begin you will need an incubator, a hygrometer for measuring humidity inside the incubator, and a regular thermometer. It takes 21 days for an egg to hatch a chick and requires close monitoring during these three weeks.

Before your eggs arrive, the incubator should warm up for one week. Read the instructions that come with the incubator carefully, as each manufacturer will have specific requirements on their product. Incubators generally come in two types: a still-air incubator or a fan-forced incubator. Air temperature in the still-air incubator should read 101.5 degrees Fahrenheit at the top of the eggs, and a fan-forced incubator temperature should read 99.5 degrees Fahrenheit. A fan-forced incubator will use a fan to evenly distribute heat throughout the incubator. A fan-forced incubator will cost more, but for small batches of eggs, a still-air incubator will give an adequate hatch.

In addition to temperature, the humidity inside the incubator is important to maintain. For the first 18 days of incubation, the humidity should range between 60 to 65 percent. For the remaining three days, the humidity should be increased to between 80 to 85 percent. It cannot be stressed enough to carefully monitor the temperature and humidity during the incubation period. Chicks will not hatch or will be unhealthy in an improperly maintained incubator.

It is essential to ensure incubator settings are correct to produce healthy chicks like these.

If you purchase eggs, you will need to settle the eggs for a day to allow the air-cell inside to return to a normal size. Those swiped from a broody hen can be directly placed inside the incubator. To settle the eggs, store them with the pointed side down at 55 to 60 degrees Fahrenheit. After they have settled, or if they have been taken from a hen, distribute the eggs evenly throughout the incubator.

Eggs must be turned every day, and using an automated egg turner is an easy way to ensure this is taken care of. If you decide not to purchase an egg turner, you can turn the eggs manually. Mark one side of the egg with an X and the other side with an O, and rotate the egg 180 degrees, which is a half-turn, five times a day. Be gentle, keeping in mind you are the egg's parent.

Rotating the eggs an odd number of times per day helps ensure that the same side of the embryo is being rotated. Turn eggs each day until day 18. After that, the chick is preparing to hatch, and should not be moved. Chicks will begin to hatch around day 21, and it should take about 24 hours for all of the eggs to hatch.

This chick worked its way through the shell.

Increase the humidity level inside the incubator and fight the urge to open the incubator door. Opening the door will allow heat and humidity to escape dropping the temperature and humidity. It will take a few hours for the incubator to return to the proper levels of heat and humidity, which may cause problems with the hatch.

Once the chicks start to break their shells, do not attempt to help them out of the shell. They need to do this on their own to develop the strength they need to survive the new world. If a chick cannot come out of its shell on its own, it will probably not be able to survive. Leave unhatched eggs in the incubator for two to three days after the first chick hatches. After that, remove the eggs and discard — as these chicks have died. If your chicks hatch wet and mushy, the humidity inside the incubator was too high. Remove the moisture from the incubator if the chicks fail to fluff up their down. Once the down is fluffed, and when they are running around the incubator, transfer them to the prepared brooder area. The chicks should be ready for transfer in a few hours after hatching.

CASE STUDY: PROPER HANDLING OF EGGS: FROM HEN TO CONSUMPTION

Phillip J. Clauer,
Poultry Extension Specialist,
Animal and Poultry Sciences

To ensure egg quality in small flocks, egg producers must learn to properly handle the eggs they produce. This article will discuss how you can ensure that your eggs will be of the highest quality and safe for consumption.

A. Layer house management

The condition of the egg that you collect is directly related to how well the flock is managed. Feeding a well-balanced ration, supplementing calcium with oyster shell, water, flock age, and health all can affect egg quality.

However, since these factors are covered in other publications, this fact sheet will place emphasis on egg quality and handling after it is laid.

1. Coop and Nest Management

   Keep the laying flock in a fenced area so they cannot hide their eggs or nest anywhere they choose. If hens are allowed to nest wherever they choose, you will not know how old eggs are or with what they have been in contact, if you can find them at all.

   Clean Environment: Keeping the layer's environment clean and dry will help keep your eggs clean. A muddy outside run, dirty or damp litter, and dirty nesting material will result in dirty, stained eggs. Clean-out the nest boxes and add deep clean litter at least every two weeks.

   Clean-out wet litter in the coop and make sure the outside run area has good drainage and is not over grazed.

   Nest Space: Supply a minimum of four nesting boxes for flocks containing 15 hens or less. For larger flocks provide one nest for every four to five hens in the flock. This will help limit egg breakage from normal traffic and daily egg laying. Make sure nests have a deep clean layer of litter to prevent breakage and help absorb waste or broken-egg material.

2. Collect Eggs Early and Often

   Most flocks will lay a majority of their eggs by 10 a.m. It is best to collect the eggs as soon as possible after they are laid. The longer the egg is allowed to stay in the nest, the more likely the egg will get dirty, broken, or will lose interior quality.

   Collecting eggs at least twice daily is advisable, especially during extreme weather temperatures.

3. Other Considerations for Layer House Management

   Rotate range areas often or allow enough area for birds in outside runs to prevent large dirt and mud areas from forming by overgrazing.

   Prevent eggs from being broken in order to minimize a hen learning to eat an egg and developing egg eating habits.

   Free choice oyster shells will help strengthen the egg shells.

Keep rats, predators, and snakes away from the hen house. They often will eat eggs and contaminate the nesting boxes and other eggs.

B. Proper Egg Cleaning and Handling

1. Collect eggs in an easy to clean container like coated wire baskets or plastic egg flats. This will prevent stains from rusted metal and contamination from other materials which are difficult to clean and disinfect.
2. Do not stack eggs too high. If collecting in baskets do not stack eggs more than five layers deep. If using plastic flats do not stack more than six flats. If you stack eggs too deep you will increase breakage.
3. Never cool eggs rapidly before they are cleaned. The egg shell will contract and pull any dirt or bacteria on the surface deep into the pores when cooled. Try to keep the temperature relatively constant until they are washed.
4. Wash eggs as soon as you collect them. This helps limit the opportunity of contamination and loss of interior quality.
5. Wash eggs with water 10 degrees warmer than the egg. This will make the egg contents swell and push the dirt away from the pores of the egg. If you have extremely dirty eggs, a mild detergent approved for washing eggs can be used. Never let eggs sit in water. Once the temperature equalizes the egg can absorb contaminants out of the water.
6. Cool and dry eggs quickly after washing. Store eggs, large end up, at 50 to 55°F and at 75 percent relative humidity. If eggs sit at room temperature (75°F) they can drop as much as one grade per day. If fertile eggs are kept at a temperature above 85°F for more than a few hours the germinal disc (embryo) can start to develop. If fertile eggs are kept above 85°F over two days the blood vessels of the embryo may become visible.

If eggs are stored properly in their own carton or other stable environment they should hold a quality of Grade A for at least four weeks.

C. Sorting and Grading Eggs

It is best that you sort the eggs before you store, sell, or consume them. The easiest way to sort eggs is to candle them with a bright light. This process can help you eliminate cracked eggs or eggs with foreign matter inside, like blood spots.

1. **How to Candle Eggs:** Hold the egg up to the candling light in a slanting position (see figure 1). You can see the air cell, the yolk, and the white. The air cell is almost always in the large end of the egg. Therefore, put the large end next to the candling light.

   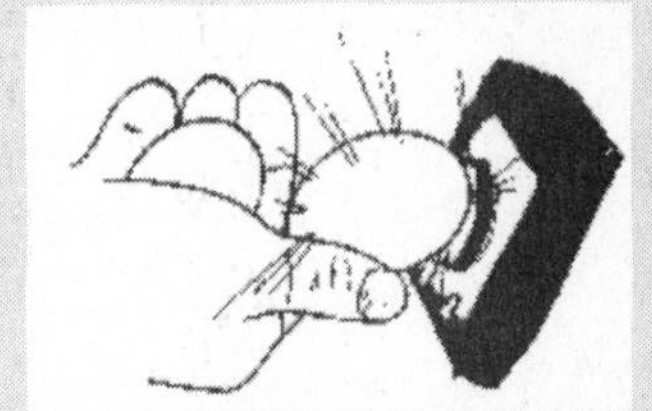

   Figure 1.

   Hold the egg between your thumb and first two fingers. Then by turning your wrist quickly, you can cause the inside of the egg to whirl. This will tell you a great deal about the yolk and white. When you are learning to candle, you will find it helpful to break and observe any eggs you are in doubt about.

2. **Identifying Cracks:** Cracked eggs will appear to have a white line somewhere on the shell. These cracks will open if you apply slight pressure to the shell. Remove cracked eggs and consume them as soon as possible or discard.

3. **USDA Grade Standard:** Use the specifications given in the table below to determine the grade of an egg by candling. Consider air cell depth, yolk outline, and albumen quality.

Quality Factor	AA Quality	A Quality	B Quality	Inedible
Air Cell	1/8 inch or less in depth	3/16 inch or less in depth	More than 3/16 inch	Doesn't apply
White	Clear, Firm	Clean, May be reasonably firm	Clean, May be weak and watery	Doesn't apply
Yolk	Outline slightly defined	Outline may be fairly well-defined	Outline clearly visible	Doesn't apply
Spots (blood or meat)	None	None	Blood or meat spots aggregating not more than 1/8" in diameter	Blood or meat spots aggregating more than 1/8" in diameter

Air Cell Depth: The depth of the air cell is the distance from its top to its bottom when the egg is held with the air cell up (*see figure 2*). In a fresh egg, the air cell is small, not more than 1/8 inch deep. As the egg ages, evaporation takes place and the air cell becomes larger and the egg is downgraded.

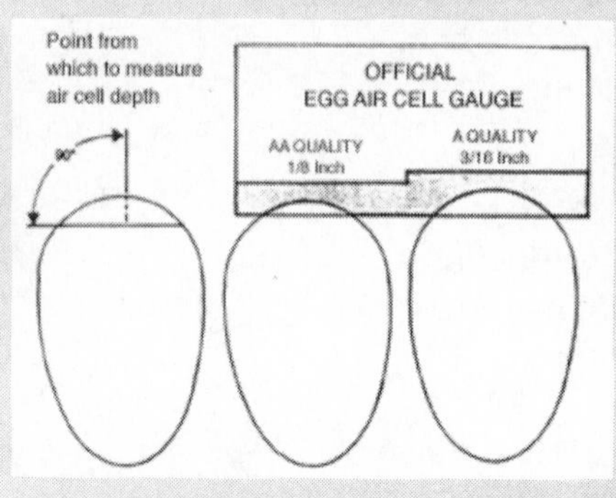

Figure 2.
Measuring air cell depth.

Yolk: The yolk of a fresh, high quality egg will be surrounded by a rather dense layer of albumen or white. Therefore, it moves only slightly away from the center of the egg when it is twirled before the candler. Because of this, yolk outline is only slightly defined in the highest quality eggs. As the albumen thins, the yolk tends to move more freely and closer to the shell. A more visible yolk when candled indicates a lower quality egg.

White or Albumen: The character and condition of the white or albumen is indicated largely by the behavior of the yolk of the egg when the egg is candled. If the yolk retains its position in the center when the egg is twirled, the white is usually firm and thick.

Eggs with blood or meat spots more than 1/8 inch in diameter are classified as inedible. Eggs with small spots collectively less than 1/8 inch in diameter should be classified as Grade B. The chalaza is distinguished from a meat spot by a bright area of refracted light that accompanies its darker shadow. Blood spot eggs can be consumed without harm, however, most people find the appearance undesirable.

4. Also remove any eggs with unusual shell shapes, textures, ridges, or thin spots on the shell if you plan to sell the eggs. These eggs are edible but break easily and are undesirable to most consumers due to appearance.

D. Storage of Eggs

1. Store eggs small end down in an egg carton to keep the air cell stable.
2. Date carton so you can use or sell the oldest eggs first and rotate your extra eggs. Try to use or sell all eggs before they are three weeks old.
3. Store eggs at 50 to 55°F and 70 to 75 percent relative humidity.
4. Never store eggs with materials that have an odor. Eggs will pick up the odors of apples, fish, onions, potatoes, and other food or chemicals with distinct odors.
5. Never hold eggs at or above room temperature or at low humidities more than necessary. Leaving eggs in a warm, dry environment will cause interior quality to drop quickly.

E. Sale of Eggs

There are no laws which prevent the sale of eggs from a home laying flock. However, you should take some basic steps to ensure that the eggs you sell have uniform quality.

1. Follow the suggestions about collection, washing, storage, and sorting above.
2. For marketing it is usually best to size the eggs. Medium, large, and extra large eggs sell best. Egg sizes are expressed in ounces per dozen.

   Small - 18 oz.
   Medium - 21 oz.
   Large - 24 oz.
   X-Large - 27 oz.
   Jumbo - 30 oz.
   Egg scales can be purchased at many farm supply stores.
3. Never sell eggs in cartons with another egg producer or store name on the carton. It is illegal to do so. Only sell eggs in generic cartons or ask your customers to bring their own carton to carry the eggs home in.

4. Most small flock producers base their prices on the current store prices in the area they live. However, many producers niche market their eggs as a specialty item and receive premium prices. If you have your birds in a fenced outside run and have one male for every ten to 15 hens in your flock, you can sell eggs at a premium as fertile, free range eggs. Brown eggs often will bring higher prices as well.

   Remember, prices will also be driven by supply and demand. If you do not have a lot of competition and have a good demand, you usually can get a higher price for the eggs you sell. It is critical that you pay attention to quality and keep a constant year-round supply for your customers. Be prepared to replace any eggs that are not satisfactory to a customer. Learn about and correct the dissatisfaction.

F. What Is the Proper Way to Cook and Handle Egg Foods?

Consumers should always keep eggs refrigerated until the eggs are used. Also, do not store eggs with other foods containing odors like onions, fish, or apples. Eggs should not be eaten raw. Pasteurized eggs should be used in recipes that call for raw eggs which are not going to be cooked (i.e. eggnog, ice cream, etc.) Eggs should not be combined and left to stand at room temperature before cooking for more than 20 minutes. Eggs should be individually cracked and immediately cooked. The USDA recommends that hot food be kept above 140°F and cold foods be kept below 40°F.

Chapter 5

BUTCHERING YOUR CHICKENS

The cycle of life means that you will need to decide what to do with your birds as they stop producing eggs or they are too old to be a part of your flock. If you are raising your birds for meat, then you already know the fate of your chickens. If your bird is a pet, and you choose to let it die a natural death, you do not need to have a plan for butchering it. Butchering a chicken can be a daunting task for some, but if you are well prepared, then the process can be quick and you will provide fresh meat for your family.

It is important to note that the process for butchering chickens is similar to the process of butchering any other breed of poultry. These directions can serve as a guide for butchering any type of bird mentioned in this book.

Many chicken owners raise their flock for meat. Chickens grow relatively quickly, and once your flock is in production, you can have plenty of meat about 12 weeks from the time you hatch your chicks. Broiler birds, such as the Cornish hen, were bred to plump up quickly and to contain adequate amounts of meat on their bones to be provided as food.

Meat birds are not as economical as you may think. The cost of raising and butchering them may be more expensive than purchasing chicken at a grocery store, but the benefit to owning your bird is that you control what it is fed and the environment it lives in. If you have not eaten organic or free-range chicken, be aware that the flavor is different than commercially produced broilers. It has more of a natural, gamey flavor. This taste may take some time to get used to, but it is healthier for you because it provides more nutrients and has undergone less processing.

You can butcher a chicken at any stage of its life, but if you wait until the bird is too old, it may have health problems. If you butcher a bird that is too young, you may not get as much meat on the bones. A bird's meat yields about 2 pounds less than the live weight of the bird. For example, an 8-pound chicken will yield about 6 pounds of meat. You may want to butcher your birds all at once, or on an as-needed basis. If you butcher them all at once, make sure you have ample freezer space to store the meat.

THE POULTRY LABEL SAYS "FRESH"

"I am shopping for a fresh chicken because I do not want the hassle of defrosting a frozen one. When should I buy it and how do I know if it is fresh? What does 'fresh' on the label really mean?"

Prior to 1997, poultry could be sold as "fresh" even if it was frozen "as solid as a block of ice." However, consumer concerns about "rock" frozen poultry being sold as "fresh" led USDA to reconsider the term "fresh" as it applies to raw whole poultry and cuts of poultry. Furthermore, national press coverage and testimonies at public hearings indicated strong interest in the term "fresh" being redefined.

After lengthy hearings, surveys, and reviews of science-based information, USDA published a "fresh" labeling rule that went into effect in December 1997. Today the definition of "fresh" is intended to meet the expectations of consumers buying poultry. Below are questions and answers about the "fresh" labeling rule and the terms "fresh" and "frozen."

Why is 26°F the lowest temperature at which poultry remains fresh?

Below 26°F, raw poultry products become firm to the touch because much of the free water is changing to ice. At 26°F, the product surface is still pliable and yields to the thumb when pressed. Most consumers consider a product to be fresh, as opposed to frozen, when it is pliable or when it is not hard to the touch.

What are the labeling requirements for frozen, raw poultry?

Raw poultry held at a temperature of 0°F or below must be labeled with a "keep frozen" handling statement.

What does the "fresh" rule mean to consumers?

For consumers, "fresh" means whole poultry and cuts have never been below 26°F. This is consistent with consumer expectations of "fresh" poultry, i.e., not hard to the touch or frozen solid. Fresh poultry should always bear a "keep refrigerated" statement.

Is there an increased microbiological safety risk associated with raw poultry that is maintained at 26°F?

No. The National Advisory Committee on the Microbiological Criteria for Foods, as well as several scientific organizations, agreed that there is no increased microbiological risk associated with raw product maintained at 40°F or below.

How should consumers handle fresh or frozen raw poultry products?

Fresh or frozen raw poultry will remain safe with proper handling and storage.

Fresh, raw poultry is kept cold during distribution to retail stores to prevent the growth of harmful bacteria and to increase its shelf life. It should be selected from a refrigerated cooler which maintains a temperature of below 40°F and above 26°F. Select fresh poultry just before checking out at the store register. Put packages in disposable plastic bags (if available) to contain any leakage that could cross-contaminate cooked foods or fresh produce.

At home, immediately place fresh raw poultry in a refrigerator that maintains 40°F or below and use it within one to two days, or freeze the poultry at 0°F or below. Frozen poultry will be safe indefinitely. For best quality, use frozen, raw whole poultry within one year, poultry parts within nine months, and giblets within four months.

Poultry may be frozen in its original packaging or repackaged. If you are freezing poultry longer than two months, you should wrap the porous store plastic packages with airtight heavy-duty foil, freezer plastic wrap or freezer bags, or freezer paper. Use freezer packaging materials or airtight freezer containers to repackage family-sized packages into smaller units.

Proper wrapping prevents "freezer burn" (drying of the surface that appears as grayish brown leathery spots on the surface of the poultry). It is caused by air reaching the surface of the food. You may cut freezer-burned portions away either before or after cooking the poultry. Heavily freezer-burned products may have to be discarded because they might be too dry or tasteless.

What is the difference in quality between fresh and frozen poultry?

Both fresh and frozen poultry are inspected by USDA's Food Safety and Inspection Service. The quality is the same. It is personal preference that determines whether you purchase fresh or frozen poultry.

What does the date on the package mean?

"Open Dating" (use of a calendar date as opposed to a code) on a food product is a date stamped on the package of a product to help the store management determine how long to display the product for sale. It is a quality date, not a safety date. "Open Dating" is found primarily on perishable foods such as meat, poultry, eggs, and dairy products. If a calendar date is used, it must express both the month and day of the month (and the year, in the case of shelf-stable and frozen products). If a calendar date is shown, immediately adjacent to the date must be a phrase explaining the meaning of that date such as "sell by" or "use before." A "sell-by" date tells the store how long to display the product for sale. You should buy the product before the date expires. A "use-by" date is the last date recommended for the use of the product while at peak quality. In both cases, the date has been determined by the food processor.

There is no uniform or universally accepted system used for "Open Dating" of food in the United States. Although dating of some foods is required by more than 20 states, there are areas of the country where much of the food supply has almost no dating.

What should you do if you find poultry that is frozen, but labeled "fresh"?

You can call the USDA Meat and Poultry Hotline and file a complaint.

Getting Ready

The first thing you should do during the butchering process is to find a location to butcher your birds. If you have children or other family members in your home that would prefer not to witness the process, select a location outside your home. Ideally, a separate shed or building with ample lighting would work well. For an indoor space, you also will need a table, water access, and some type of drainage for waste. If you opt to butcher outdoors, you may want a temporary screen or fence for discretion, but the choice is yours.

The following is a checklist of items you will need prior to butchering your birds.

Checklist: Items you will need

- Ax, meat cleaver, or large sharp knife. You will also need a knife sharpener. Country Horizons offers these for sale, in addition to other poultry and farming products (**www.countryhorizons.net**).

- Table with cutting-board top. Two companies that specialize in a variety of cutting boards for any purpose are Butcher Block Company (**www.butcherblockco.com**) and AWP

Butcher Block (**www.awpbutcherblock.com**). The size of the table can depend your preference. A small butcher-block table is about 28 inches wide by 24 inches deep and 33 inches tall. The AWP Butcher Block, Inc. offers this size and the next size larger, which is 48 inches wide by 24 inches deep and 36 inches tall. Find the size that meets your needs, and be sure it is treated to prevent microorganisms and germs from inhabiting the block of wood.

- Rubber gloves. You can find these at most grocery stores and department stores or Rubbermaid's online store (**www.rubbermaid.com**).
- Rubber apron. You can find these at a sporting goods store or in the garden department of large department stores.
- Large pot to boil water. This pot should be large enough to fit your largest bird.
- Stove or propane burner large enough to fit the pot.
- Kill cone or hooks or large nails. Find kill cones at feed stores and hardware stores. Sure Hatch is a company that sells kill cones and other poultry items (**www.surehatch.com**).
- Trash receptacle.
- Plastic tubs to ice down the birds after butchering.
- Sink or hose with running water.
- Boning knife or butcher knife. These are sold at sporting goods stores, kitchen supply stores, and larger department stores.

- Rags or paper towels.
- Cleaning spray, soap, or disinfectant.
- Freezer bags. The size will depend on the size of the bird you want to freeze.
- Freezer space to store your birds.
- Work table to debone and carve the carcass.

Before you butcher your chicken, you will need to put on a rubber apron, gloves, and protective eyewear. This will keep you clean and safe from cuts and scratches. Next, you will need your equipment. Some people kill chickens with their hands; others prefer to use a sharp knife or ax. Always make sure the blade is sharp. The table you use will be on the receiving end of the blade, so it should have a chopping-block top or be made of wood. Use a tree stump if you are outdoors.

A piece of cone-shaped metal or plastic, known as a killing cone, can be purchased from sporting goods stores or online. The cone slides over the chicken's head and is inverted either to kill the chicken or let the blood drain from a dead chicken. If you do not have a cone, you can hang the carcass upside down from a nail or hook for the same effect. Hang them at least 3 feet off the ground, over a bucket to collect the blood. You will need a trash receptacle for the waste.

Some hanging poultry, waiting to be butchered.

Butchering is best done before daylight. The chickens will be sleeping in their hen house, and it will be easier to pick them up and bring them to the slaughter area. The dark helps keep the birds calm, and the quietness of the morning hours will keep them from being excessively stressed. Birds that are less stressed will taste better because they have fewer hormones running through their body. Also, they bleed cleaner, making the butchering less messy.

As mentioned earlier, do not feed the birds the night before, and provide little if any water. The birds will be easier to clean if their digestive tracts are empty. Also, keep their coop dark. When entering the coop in the morning, be quiet and calm as you collect your birds. Ideally, you want to have the slaughter finished before sunrise. This will allow time for the bird to drain and be butchered for dinnertime.

Methods of Killing Your Chickens

There are several different ways to kill your chickens. Whichever method you choose to use, your priority should be to provide a fast, painless death for the chicken. The first time you butcher an animal, it may be difficult for you. This is natural, especially if you raise the chickens and become attached to them. A humane death is an honorable ending for any bird.

Without a knife or ax

One way to kill a chicken is to wring its neck. Pick up the chicken and hold it upside down with one hand. Slide your free hand down the chicken's neck to just below the bird's head and take hold. Grasp the neck firmly and jerk it down and back up again with a twist to break the chicken's neck. Pull hard, but be aware that if you pull too hard, you may yank its head off.

After killing the chicken, you will need to cut off its head and turn it upside down in a kill cone or hang it on a nail or hook to let the blood drain. Dispose of the blood by pouring it down the sink or drainage system and flush with lots of water. Clean and disinfect the entire area.

With a knife or ax

First, put on your rubber apron and then make sure your knife or ax is sharp. Dull blades will only cause the animal to suffer and will not get the job done. Get your knife professionally sharpened or use a knife sharpener.

If you are using an ax, you will need a tree stump or table with a top that can handle the blade slicing into it. The surface you choose should be low and give you enough room to swing your ax up and then bring it down on the bird's neck. Also, you will need to hammer two nails into the stump or run a wire across the area. The chicken's neck will slide between the nails or under the wire to hold it in place. The nails only need to be as far apart as a chicken's neck, and the wire only has to be loose enough to slide the chicken's head underneath it.

Hold the chicken upside down, with your ax in one hand and the bird in the other. Slide the chicken's head between the nails or under the wire. Do this quickly. Pull back on the bird's legs slightly so the neck is stretched out. In one swift move, strike the ax down on the chicken's neck, making a clean, quick cut. Hold on to the feet because you do not want to let the bird go, as it will move. The body will still move in those first few moments after the kill because of residual nervous energy. Do not let the bird move after killing it because the experience can be traumatizing to a beginner.

Kill cones make slaughtering a bird with a knife easier. Hang the cones in your slaughterhouse or somewhere you plan to do the killing. Put your bird in the cone, with the small opening on the bottom and the large opening on the top. Your bird will be upside down. If you do not have a kill cone, tie the birds together and hang them from a nail or hook that is at least 3 feet from the ground. Kill the bird within seconds of putting it in the cone or hanging it from the nail. It is not humane to let it just hang there.

Once the bird is hanging, stretch its neck a little and then slit the throat with a sharp knife. You can either perform one clean, quick cut to remove the chicken's head, or you can make a slit to only drain the blood. If you do not immediately take the head off, the bird may feel some distress for a few moments until it bleeds out. Once the blood is drained, then you can cut the head off. Have buckets beneath the cones to catch the blood. Let the chickens hang until all of the blood has run out of their bodies.

These headless chickens are on the butchering block ready to be cut.

Butchering is a messy task. This is why some owners prefer to do a group slaughter, which entails killing more than one bird in your flock at the same time. The process is still the same; however, you need more kill cones to hang the birds upside down. Gather your birds one at a time. After you have slit the throat or removed the head, hang your bird in the kill cone, then get another bird and do the same. After you have your desired amount of birds slaughtered, proceed with the butchering.

Processing the Carcass

The next step is to pluck the chickens. To do this, put on your rubber gloves if you have not already. Have a large pot with scalding hot water ready to soak the chickens to kill germs and clean them. The water should be about 140 degrees Fahrenheit, which you can test with a candy thermometer or deep-fryer thermometer. Hold the chicken carcass by the feet and dip it into the pot for about ten to 15 seconds. Pull the bird out and try pulling one of the feathers. If it comes out easily, the chicken is ready to be plucked. If the feathers do not come out easily, dip the chicken again. Keeping the bird in the water for too long or having the temperature too high will cook the skin, so monitor the process. If the chicken is partially cooked, cool it immediately or discard it so bacteria do not grow in the meat. Warm temperatures are the perfect breeding ground for bacteria.

At this point, when the feathers are ready to come out, dip the bird into an ice bath set up in one of the plastic tubs, to help prevent tears in the skin. This is not necessary, but it may make plucking easier and kill heat-resistant bacteria. Now, pull out the feathers in the direction they were growing. This may seem time-consuming, but it is faster than some of the automated plucking machines available commercially. Once you get the hang of it, the process will move quicker.

After plucking, examine your bird. Make sure the flesh does not have any sign of disease. If you have a small flock, you should have been monitoring your birds all along. If your flock is large, it may be hard to keep track of each one, especially if they have hidden lesions or bumps on their bodies. If you come across a chicken with abscesses or lumps filled with pus, do not eat the bird.

Other problems to look for are sores or open wounds and tumors. If you find them on a chicken, discard the carcass. These sores and wounds can be signs of something toxic for you if consumed. When in doubt, throw it out. Throw out the carcasses if the butchered birds were left for more than an hour at temperatures over 40 degrees Fahrenheit. Do not eat birds that were found dead.

Inedible and Removable parts

- Neck
- Oil gland
- Crop, esophagus, and trachea tubes
- Tail
- Spine
- Organs
- Feet

Now that you have inspected your plucked bird, take the carcass to the worktable and get your knife out. Remove the chicken's head if you have not done so already. If you want to save the neck, carefully remove the esophagus and trachea tubes. If you do not want to save the neck, slice it off near the body and discard it.

Cut the feet off, slicing through the cartilage above the foot at the first joint. This is easier to cut through than the bone. Lift and then slam down a meat cleaver or sharp knife to make a clean cut. Discard the feet unless you like to cook with them. If you do want to keep them, put them in a storage container and refrigerate. They can be deep-fried or made into a soup.

At the bottom of the spine, you will see a yellow spot or yellow bump near the tail. It is the oil gland. Lay the bird breast-side down. Lop off the tail at the spine and throw it away. Or, if you prefer to leave the tail on, take your knife and slice under the oil gland, down and past the tail, to cut out the gland and bypass the tail. It is not advisable to leave the oil gland in because it gives the meat a bitter taste if you do.

Flip the bird over onto its back, and cut into it above the vent. Your objective is to make a small hole in the carcass to remove the organs. Do not cut too wide or too deep, but the hole should be wide enough to fit your hand into. Stick your fingers in and pull apart the skin. If you find bird feces, wash them out. Take the carcass to the sink or to a running hose and, while holding your bird, flush out the feces. The water should run down the vent side, not across the whole body, so it does not get contaminated. You may even want to use a mild dish detergent to wash off any affected areas. Rinse completely.

Once the bird is clean, wash any contaminated areas on your table. If you plan on cutting the bird into pieces, use kitchen shears to slice through its back and remove all of the organs. Otherwise, you will need to insert your hand inside the bird. Place your hand inside the carcass and move your arm up until you reach the bird's neck. Spread out your fingers, as much as you can, and rake your hands down the inside of the carcass to pull the organs out. Do this gently. Do not grab, as you may break open some of the organs inside the bird. Once you have pulled the organs down to the vent, scoop them out and toss them into the trash.

At this point, check the chicken's liver, which should be reddish-brown. If it is pale or discolored, it is probably diseased, and you should discard the chicken. If you want to save the heart, liver, and gizzards, sort through the internal mass that you just removed and separate the organs you want to save. Once you have done that, double-check to make sure the cavity is clear of debris. Take the chicken back to the sink or water source and rinse it inside and out. Refrigerate your bird as soon as possible.

Spray down the cutting area with disinfectant or bleach cleaner, and use rags or paper towels to dry the area well. Spray down the area a second time and repeat the process. Be thorough in your cleaning, as harmful bacteria can live on countertops and surface areas. Clean the kill cones by soaking them in warm bleach water to disinfect, then rinse them well. Wash your rubber gloves and aprons by spraying them with bleach and rinsing. Clean your hands every time you come into contact with the carcass or pieces of the bird. Good sanitation is important for you and your family's safety.

Cutting your chicken

After your bird is plucked, cleaned, and processed, it is ready to cut. You may not have the skills of a professional butcher, but you can cut your chicken similar to the way you would find it packaged. The simplest way to cut your chicken is to begin with cutting the chicken in half lengthwise, which means down the breast, and then cutting the parts away width wise. You will need a very sharp chef's knife and a pair of very sharp kitchen shears. In just a few simple steps, you can have your chicken completely cut up.

- **Step one** – With your kitchen shears, cut the chicken in half along the breastbone. You need to make sure that your kitchen shears are very sharp in order to accomplish this.

- **Step two** – Flip the chicken over and cut along both sides of the bone to remove it from the breasts. You can either discard the bone or save it, storing it in the freezer so you can use it to make your own stock.

This man is selling cut up chicken in a market.

- **Step three** – Now you should be staring at two halves of the chicken. The next thing to do is flip one of the halves over so that it is laying skin side up. Then, using your sharp chef's knife, cut halfway between the wing and the leg. If you are having difficulty cutting it, then you can place the knife in the spot that you want to cut it and push down on it with your other hand. Repeat this same process with the other half of the chicken. If you want to, you can cut the remaining pieces in half a second time.

Professional Butchering

You may know right from the start that you do not want to butcher your birds. If no one in your household wants to slaughter your birds, you will need to find someone to do it for you, such as a friend or fellow chicken-owner. You could also have a professional butcher handle the slaughter for you.

In the early 1900s, butcher shops were common on city streets. There was an art to butchering. In cities such as New York and Los Angeles, butchers may still have storefronts. Today, though, most grocery chains have their own meat departments inside the stores, and machines now take the place of professional butchers.

These whole chickens are for sale in a butcher shop.

Your best resource for finding a local butcher is to search the Internet or the phone book. You may have to include surrounding areas in your search if your area does not have a butcher, but you will likely have the most luck in an Internet search. If you still cannot find a butcher, try advertising in the local classifieds under the "wanted" section. Or go to your local grocer and ask someone who works in the meat department whether he or she can do it for you or can recommend someone who can.

A knowledgeable butcher will know how to cut any type of meat. Chicken is a common meat, and any butcher should be able to cut it up for you. Butchers should know a healthy bird or carcass when they see it. They should not be willing to cut up poultry that is sickly or tainted. A good butcher will know how to make clean cuts, and will be able to distinguish prime cuts of meat from lesser cuts. Most butchers have been an apprentice or have on-the-job training. Your butcher should be able to discuss the parts of the bird with you in detail and have a clean environment in which your bird is prepared.

Chicken handling tips

No matter if you process and cut your chicken yourself or find a butcher to do it, the result is the same – you have fresh meat for you and your family. Storing and handling raw chicken is very important when it comes to the health of your family as knowing the proper food handling methods will keep your family from getting salmonella poisoning. Here are some tips to follow to ensure you keep your family and kitchen supplies safe:

- Always use a solid plastic cutting board when you are working with raw chicken. Immediately after each use and before you use it for anything else, you must scrub the board with hot, soapy water. If you are having a moment of doubt in your ability to properly clean the cutting board, run it through the dishwasher. It is alright to use a wooden cutting board, but make sure to scrub it in water that is hotter than you can handle. Make sure to use a pair of rubber gloves so that you do not burn your hands.

- You must scrub any knife or utensil that comes in contact with the raw chicken in hot, soapy water immediately after it is used.

- Never defrost a chicken on your kitchen counter; this just breeds bacteria onto your surface top. To properly defrost a chicken, you should place it in the refrigerator in its original packaging and in a bowl of very cold water that needs to be changed every 45 minutes. You can also defrost your chicken in the microwave.

- It is utterly important that you discard any type of marinade that you used to coat the chicken in during its storage time. Do not ever baste a chicken with the cold marinade it was stored in.

- Do not ever serve your family "rare" chicken, because there is no such thing. A chicken has to be fully cooked, which entails that its juices run clear and, more importantly, that the internal temperature reads 165 degrees Fahrenheit when a meat thermometer is inserted into the meat.

- Sometimes, you will find a recipe that calls for you to pat dry a chicken before you cook it. Do this using a piece of paper towel; never use a dish towel to pat the chicken dry.

- If you are going to handle the chicken with your bare hands, always remember to vigorously wash your hands in hot, soapy water before you touch anything else. Touching the raw chicken and then touching a different substance, such as butter, vegetables, or utensils, can spread the chicken's bacteria to other things without you knowing it, which will in turn cause the next person who uses or eats it to be sick. Whenever possible, use gloves to handle the raw chicken, then take them off immediately after you finish handling the chicken.

New Homes for Chicks and Chickens

You may decide that butchering your birds is not for you, or at some point you may decide that you no longer want to keep birds. What should you do? One option is giving your birds to a local farmer. Experienced farmers might be happy to take your birds from you, providing they are healthy and in good condition.

In case you find yourself in a predicament where you need to get rid of your birds, keep any health records and visits to the vet on record to give to the new owner. This way, the owner can continue to care for the birds properly. Having this information will help you place your birds more quickly.

Chicks are adorable. If you plan to sell chicks, make sure your buyers are prepared to care for them properly and are not buying them for novelty.

You also can try to sell your birds. Be honest about their ages and health histories. List your birds for sale in the local classified ads and on online websites:

- CC.CC (**www.cc.cc**) allows you to set up your own domain for free.
- Plaza.Net (**www.ec.plaza.net**) allows you to post your chicken eggs for sale.
- Backpage.com (**www.backpage.com**) lets you post items for sale for free. First, locate your city on the homepage, then find the appropriate post to sell your birds. There is a farm and garden section and a pet section.
- Freeadforum.com (**www.freeadforum.com**) is a website that will allow you to post and sell your chicks and eggs for free.

Some neighborhoods have exchange programs where you can offer your birds and supplies to someone who has something to barter with you. The social networking website Facebook (**www.facebook.com**) also has exchange programs that may help you find a new place for your flock and get something for yourself as well.

Even if you no longer want your birds, do not neglect them. It may take a little time, but you will find a solution. Problems will just compound if you do not care for your poultry and they get sick or die off. Disease can create havoc. If time is of the essence, contact your local animal shelter or animal control officer and explain your situation. They should be able to assist you or at least put you in contact with someone who can help.

Natural life span of chickens

The life span of chickens is between five and seven years, although some breeds can live longer. There is always the rare situation where an animal may live much longer than their life expectancy due to a healthy and safe environment and probably good genetics as well. Chickens can provide many substantial years of companionship and food for you. Their life with you can be rewarding and fulfilling for both you and your birds.

The life cycle of the chicken starts with a fertilized egg. The hen sits on her eggs to hatch them. Twenty-one days later, baby chicks emerge from the eggs. The chick's body is covered with soft down feathers. Chicks can walk right away. Within the first four weeks of life, chicks will grow more feathers and be able to eat bugs, worms, and seeds.

Chicks are full grown at 6 months of age, and most hens begin laying eggs at 4 or 5 months of age. Different breeds may vary on their maturity level, and some hens may not produce at all due to health problems or other outside factors. The cycle of life begins again with each new fertilized egg. Chickens constantly give back to their owners and you will be pleased with your return

Raising chickens is a rewarding hobby.

whether it is their companionship, their eggs, their meat, or their reproduction of baby chicks. You will have a more bountiful, sustainable existence thanks to your new flock of feathered friends.

Pulling it all together

Now that you have the information you need to start a flock, the next step is to get in there and do it. Doing is the best teaching tool in raising chickens or other poultry. You have done yourself and your future flock a great favor by reading how to raise and take care of chickens. The next step is to do your prep work for your new arrivals. Most importantly: Have fun!

CASE STUDY: FAMILY FOWL

Jeff Nardello
JNJ Natures Way Farm
Sebeka, MN
www.jnjnaturesswayorganicfarm.com

When a former law enforcer wanted to provide healthy meat for his family, he chose to stop buying organic and start growing it.

"We decided to raise chickens and turkeys because we do not want the garbage presented in grocery stores for our family," said Jeff Nardello of Sebeska, Minnesota.

Nardello and his family specialize in raising small flocks of Cornish cross chickens and white broad-breasted turkeys in addition to beef, pork, and lamb. Nardello said they were tired of consuming the antibiotics and other chemicals most mass-producers subject their poultry to. Instead, they feed their birds natural feed and allow them to graze on fresh grass and bugs.

Though the Nardellos do not need to devote much time to their flocks — only about nine to ten hours per week — they do not make much profit on the fruits of their birds.

"In our area, it is hard to sell organic eggs for a profit, as the consumer does not know that an organic egg is worth more than the egg they can buy at the store for 99 cents," Nardello said.

They market their beef, pork, and turkey on their website and also sell eggs and chicken meat locally. He and his family enjoy the work, as it is usually short-lived and rewarding.

"Each animal is raised with respect and care," Nardello said. "Life of any kind is special, and we like to see the animals thrive under our care."

The end product, he said, is the best part of the process. Having fresh, safe meat for his family is very healthy. The family works to keep their farm disease-free. Salmonella is especially a concern, but Nardello said it is easily avoided with proper hand washing.

Another great advantage is that it is something the whole family can get involved in and learn something from. Nardello's 8-year-old son decided to forego the usual lemonade route and sell eggs instead by raising chickens himself.

"Our son would hold the babies, and they would not really bother him once they were used to him and him them," Nardello said. "It was a great experience for him."

If your family wants to start its own small flock for organic eggs, Nardello suggests between 20 and 25 birds for one and a half dozen eggs per day, on average. If a family wanted to produce meat as well, he suggests about 50, since the average family eats about 25 chickens a year.

You are now equipped with the knowledge to raise a healthy flock of chickens.

Part 2

TURKEYS

Turkeys are a bird that beginners can raise without many issues. The breeds are varied in coloring, behavior, and uses, so anyone interested in poultry can find a breed that appeals to them. Turkeys can provide your family with meat and eggs, and if the market is right, you can also make a profit from raising your birds. Consider raising turkeys if you want to help in conservation efforts, have a unique pet, supplement your income, or take up a new hobby that provides food for your family.

Chapter 6
TURKEYS

Another poultry choice aside from chickens is turkeys. Turkeys might make a great addition to your farmstead while adding to your family's freezer or as a potential income generator. Turkeys can be trickier to raise than chickens because the young – called poults – are more fragile than chicks. But this can be overcome by reading all you can about these big, beautiful birds.

History

The turkey is native to North America. The common wild turkey (*Meleagris gallopavo*) roamed through the continent from Canada to northern Mexico. The species is divided into six subspecies, each adapted to its particular geographical area. These subspecies are still found in the wild today.

- The eastern wild-turkey inhabited the eastern half of the United States.

- The Florida wild-turkey, or Osceola, roamed Florida.
- Merriam's wild-turkey inhabited the mountain regions of western United States.
- The Rio Grande wild-turkey inhabited the south-central plains states and northeastern Mexico.
- Gould's wild-turkey was found in northwestern Mexico and parts of southern Arizona and New Mexico.
- The Mexican wild-turkey inhabited southern Mexico.

Central America had its own subspecies: the Ocellated turkey (*Meleagris ocellata*). It was, and still is, found on the Yucatan Peninsula of southeastern Mexico and may be a separate species from the other wild turkeys. The male has a blue head with random round, pink growths. Its bronze-green feathers have a metallic sheen. The wing tips are white and the tail tips are a blue-bronze. The tail feathers also have peacock-like spots. It does not gobble like other turkeys; instead it makes a whistling sound.

Although the exact date of domestication of the turkey is unknown, they were definitely used as meat producers when the Aztecs were taken over by the Spanish. The Aztecs used the meat and eggs for food and the feathers for decoration and clothing. While they used the meat as a protein source and the feathers for decoration, it is thought that religious reasons were the main focus of domesticating the bird.

By the time Spanish explorers arrived in North America, the turkey was fully domesticated. Montezuma even presented Cortes with six golden turkey statues. These explorers brought the turkey from North America to Europe where it became an instant hit. In 1511 King Ferdinand of Spain ordered Spanish ships arriving from North America to return with turkeys. In Europe, breeding programs of these wild turkeys developed many varieties of domesticated turkeys. The Europeans gave the bird the name "turkey" because they mistakenly believed that the bird came from Turkey. When colonists from Britain settled in North America, they brought the domesticated version to the colonies to be used as a food source, but the wild turkey was still alive and present. It was a well-regarded bird; Benjamin Franklin even suggested it as a candidate to become America's national bird. In a letter to his daughter, Ben Franklin wrote:

> "*For my own part I wish the Bald Eagle had not been chosen the Representative of our Country. He is a Bird of bad moral Character. He does not get his Living honestly. You may have seen him perched on some dead Tree near the River, where, too lazy to fish for himself, he watches the Labour of the Fishing Hawk; and when that diligent Bird has at length taken a Fish, and is bearing it to his Nest for the Support of his Mate and young Ones, the Bald Eagle pursues him and takes it from him.*
>
> *With all this Injustice, he is never in good Case but like those among Men who live by Sharping & Robbing he is generally poor and often very lousy. Besides he is*

a rank Coward: The little king Bird not bigger than a Sparrow attacks him boldly and drives him out of the District. He is therefore by no means a proper Emblem for the brave and honest Cincinnati of America who have driven all the king birds from our Country...

I am on this account not displeased that the Figure is not known as a Bald Eagle, but looks more like a Turkey. For the Truth the Turkey is in Comparison a much more respectable Bird, and withal a true original Native of America... He is besides, though a little vain & silly, a Bird of Courage, and would not hesitate to attack a Grenadier of the British Guards who should presume to invade his Farm Yard with a red Coat on."

Maybe Ben was onto something.

The modern turkey

This wild turkey is equipped to survive a cold, snowy winter.

The modern turkey has undergone tremendous change due to the commercial turkey industry. Unlike their wild brethren, domesticated turkeys are much heavier and have a greater muscle mass. Through selective breeding, commercial strains of turkeys have massive muscles on a normal-sized frame. Their muscles have become so large that most commercial strains can no longer breed naturally. The tom (male) is unable to perform the act of mating with the hen (female), so eggs remain infertile unless human intervention occurs thorough artificial insemination. Most commercially bred hens are mated through artificial insemination.

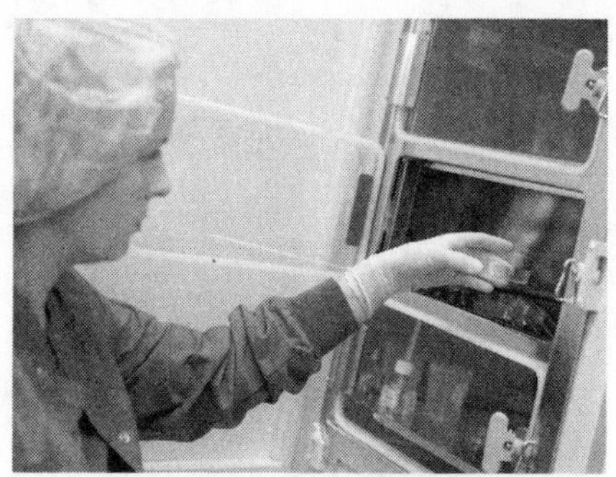

An embryologist placing a sample into an incubator.

Much like the commercial chicken industry, a typical large commercial-turkey farmer contracts with the turkey processing company to raise a certain number of birds. Thousands of turkeys are raised in large, long sheds. Often, commercially raised turkeys live their entire lives in these indoor spaces. A few growers do grow substantial numbers of turkeys on pasture or free-range situations, but it can be hard to control death losses due to disease and poor weather with large numbers of turkeys.

Turkey Anatomy

Turkeys have some unique body parts that characterize the species. Their heads are featherless, and have a caruncle, a brightly colored growth on the head and throat. This skin is carunculated and is indicative of a turkey's mood. It is normally reddish-pink but will become bright red if a turkey is upset. This can help you gauge the mood of these big birds when you work with them. During courtship, this skin will also turn bright red. The snood is a flap of skin beginning at the base of the top of the beak. In female turkeys this flap is just a little knob but in males it can stretch out to hang well beyond the tip of the beak when they are excited. The snood is the same color as the caruncle and likewise will turn bright red if the turkey is upset or courting. The wattles on a turkey hang right under the turkey's chin. It extends out from the carunculated skin and is a thin flap. The wattle will also turn red during courtship or if the turkey is agitated.

You can see the caruncle, snood, and wattles of this adult, male turkey.

Turkeys have a long, fan-shaped tail, and male turkeys will frequently strut about the barnyard with the tail fanned out. Each foot of a turkey has three toes and males will have a spur, a long hard growth on the back of the lower legs. The spur grows sharper and longer as the turkey ages and can be a formidable weapon. Turkeys, like all birds, have pinfeathers, which are also called blood feathers because these feathers contain a blood vein which will bleed profusely if the feather is pulled. Pinfeathers generally have a plastic-looking sheath around them. In meat-producing turkeys, lighter pinfeathers make processing the carcass less difficult. Male turkeys usually have a beard, a tuft of coarse hair that grows from the center of the breast. Females will occasionally have a beard, but it will be much thinner and shorter than the male turkey's beard. A turkey's age can be estimated by beard length. The beard grows at a rate of 4 to 5 inches a year and does not stop growing. However, by the time the beard is 9 to 10 inches long the tips start to wear out. The beard will also become thicker and wider as the bird ages.

Turkey Breeds

The majority of commercial growers raise the variety of turkey called the broad-breasted white. This is not an official breed; rather it is a commercial variety of the bronze turkey breed, which is described in the heritage breed section. The commercial varieties of the broad-breasted white can be purchased from turkey hatcheries, and each hatchery will have their own version of the commercial broad-breasted bronze. These birds, if they have a constant source of feed, can grow to market weight of more than 30 pounds in 18 weeks.

Heritage breeds

Heritage breeds are varieties of turkey that have historic characteristics but are no longer commercially raised for consumption. Heritage breeds are in danger of extinction as their total population numbers are low. Some people prefer these breeds for the table, claiming they have a better taste than the commercial varieties. They will take longer to grow to market weight than the commercial breeds, usually around 24 to 30 weeks.

Beltsville small white

Tom: 23 pounds
Hen: 13 pounds

As described by its breed name, this is a small, white turkey that was developed by the U.S. Department of Agriculture Research Station in Beltsville, Maryland in the 1930s and 1940s. Scientists used a number of breeds to make the Beltsville small white turkey: white Holland, white Austrian, bronze, and black among them. Its white feathering means there are less visible pinfeathers than in the colored turkey breeds and its smaller frame is popular with those consumers wanting a smaller table bird. It was popular during the 1950s and accounted for more than a quarter of the turkey sales during this time. With the advent of large-scale turkey production, the breed almost became extinct in the 1970s.

Black

Tom: 33 pounds
Hen: 18 pounds

This breed has greenish-black feathers. It is rare in North America, but it is bred in Europe where it is considered to be a tasty table bird. Both Spain and England claim its development. It was developed from the first turkeys brought back from North America in the 1500s. It was among the first domestic turkeys to return to North America with the settlers. There are different varieties of the black turkey. The Spanish black is solid black including feathers, legs, and toes. The Norfolk black from England can have white-tipped feathers. Their legs and toes can also be pink.

Blue slate

Tom: 30 pounds
Hen: 18 pounds

This turkey is also called lavender due to its color. This breed is further classified based upon two color patterns: Blues are a solid, dull, gray-blue, and slates are an ashy blue color with specks of black. Blue turkeys bred to blue turkeys can result in three different colors: slate, blue, and black, which come from its heritage from black turkeys. A slate turkey bred to another slate turkey will always result in slate-colored offspring. This breed is listed as critical for extinction, as there are few of these turkeys left in the world.

Bourbon red

Tom: 33 pounds
Hen: 18 pounds

As its name suggests, this handsome bird has rich red feathers laced with black with white-tipped wing and tail fans. The breed was developed in Kentucky and Pennsylvania from crossbreeding with the bronze, buff, and white Holland turkeys. It was an important commercial turkey in the 1930s and 1940s. The bourbon red turkey still remains popular with small-scale turkey growers. It is an excellent forager and has a meaty breast. The carcass picks clean with light-colored pinfeathers.

Bronze

Tom: 45 pounds
Hen: 19 pounds

This breed was developed in the United States in the 1700s from crossbreeding turkeys brought from Europe with the eastern wild turkey. The feathers are bronze-colored with copper or blue tints. The wing feathers have stippling of white at the ends. The broad-breasted bronze variety has massive breast muscles and is no longer able to breed naturally. Prior to the broad-breasted white turkey becoming the popular commercial turkey, this breed was the popular commercial turkey. There is an unimproved variety (the standard bronze) that is true to its heritage. It was in danger of dying out, but it is now making a comeback with backyard breeders. It has essentially the same coloring pattern as the broad-breasted bronze.

Narragansett

Tom: 33 pounds
Hen: 18 pounds

This turkey has gray, tan, black, and white feathers in a similar patter to the bronze turkey. It was developed in the eastern United States by crosses of turkeys from Europe and the eastern wild-turkey in the 1700s. It is named for the Narragansett Bay in Rhode Island. The Narragansett turkey has dark feathers with a steel-gray tinge. The Narragansett turkey has a calm disposition, and the hens are broody.

Royal palm

Tom: 22 pounds
Hen: 12 pounds

Unlike most turkey breeds, the royal palm turkey was bred more for its unique look than for its meat production. It has white feathers with black feathers stippling the breast and covering a wide swath along the base of the tail. The tail feathers have a wide band of black a few inches from the tail tips. They are good foragers and produce a small carcass perfect for a smaller family.

White Holland

Tom: 33 pounds
Hen: 18 pounds

As the name suggests, this is a white-feathered turkey that originated in Europe. Dutch and Austrians became attracted to its unique all-white feathers. The turkey has a red or blue head and a black beard. Its eyes can be either brown or blue. The white Holland was an important commercial bird in the United States during the early 1900s, but its numbers have declined to the point that it is currently threatened with extinction.

Broad-breasted white

Tom: 50 pounds
Hen: 36 pounds

This is the turkey that developed the large commercial breeds used today. It is the most common turkey sold for turkey consumption in the United States. The breed was developed from crossing the broad-breasted bronze and the white Holland breed. Due to its massive size, they tend to suffer from cardiac and leg problems. The turkey has white feathers but is larger than its ancestor, the white Holland.

Jersey buff

Tom: 25 pounds
Hen: 14 pounds

The Jersey buff turkey has an unclear origin, but it is believed to have been developed in the mid-Atlantic states. It is a tan-colored bird with white wing and tail tips. The light coloring leaves a clean carcass with light-colored pinfeathers. The breed is hard to breed true to color, but researchers at the New Jersey Agricultural Experiment Station did attempt to bring the breed back in the 1940s. The hens are good egg-producers.

Auburn

Tom: 35 pounds
Hen: 19 pounds

The Auburn, also known as the light brown, is an old variety of turkey. There are records dating from the late 1700s in Pennsylvania regarding the transportation of Auburn turkeys to market. During this time, the Auburn was one of the largest birds available. This breed has reddish-brown feathers when mature. Often, males and females will have different coloring — some with lighter feathers and some with darker feathers. Currently, the breed is one of the rarest and is in extreme danger of becoming extinct due to low-breeding flock numbers. They can reproduce naturally.

Chocolate

Tom: 33 pounds
Hen: 18 pounds

The chocolate turkey is named for the chocolate color of its feathers, feet, and legs. It used to be a common turkey in France. In the United States, the chocolate turkey was popular in the pre-Civil War southern states, but it lost popularity during the Civil War and never regained its previous status. Today, most of the chocolate turkeys sold will have genes from the Narragansett and bronze turkeys. Any white or bronze discoloration of the feathers is considered a breed fault.

White midget

Tom: 18 pounds
Hen: 10 pounds

This small turkey was developed in the 1950s in Massachusetts. They are quite similar to the Beltsville white turkey and are frequently confused with that breed. The white midget almost died out, but the University of Wisconsin kept a small flock of the turkeys. Now, hatcheries have become active breeders of this small, family-table friendly bird. The bird is white and is about the size of the larger chicken breeds.

Purchasing heritage turkeys

Hatcheries that sell heritage breeds of turkeys will have a lot of information regarding specific breeds. Most of these hatcheries are run by enthusiastic turkey boosters and are proud of their part in preserving the breeds they raise. Here are some websites for hatcheries that sell heritage turkeys:

- Porter's Rare Heritage Turkeys: **www.porterturkeys.com**
- Cackle Hatchery: **www.cacklehatchery.com**
- Stromberg's: **www.strombergschickens.com**
- Murray McMurray Hatchery: **www.mcmurrayhatchery.com**

Heritage Breed Preservation

For a domesticated turkey to be "heritage," it must contain some historical characteristics that modern commercial turkeys lack. The American Livestock Breeds Conservancy has determined the following criteria in order to determine if a turkey is a heritage turkey:

- The heritage turkey must mate naturally, even when the bird is bred for genetic maintenance. A heritage turkey should have fertility rates of 70 to 80 percent.
- The sexual productivity of a heritage turkey must last for several years; hens should lay eggs for five to seven years and breeding toms should breed for three to five years.
- The heritage turkey must have a slow- to- moderate growth rate – usually 36 to 28 weeks. A slower growth rate allows the bird's body time to develop a strong skeleton and healthy organs.

As a turkey producer, you may find a special niche in breeding and maintaining a flock of heritage turkeys. Since the broad-breasted white and the broad-breasted bronze turkeys require artificial insemination to reproduce, most backyard and small-flock turkey owners will not have the skill to perform this procedure. A heritage turkey will be able to breed naturally and will not have the heart and leg problems associated with the two commercial breeds of turkeys.

The Heritage Turkey Foundation (**www.heritageturkeyfoundation.org**) is an organization founded to protect the surviving heritage turkey strains and reintroduce them to the American marketplace. Their focus is to encourage farmers to raise heritage turkey breeds and promote the turkeys in the marketplace and to consumers. First, these breeds should be preserved for genetic reasons. Commercial turkeys are susceptible to health problems that could be catastrophic for the entire industry. The gene pool is small, making one disease more likely to wipe out an entire population of turkeys. Second, preserving heritage turkeys can also benefit consumers. A slow growth rate gives heritage breeds a flavor not available from a commercially produced turkey. Although this meat may be more expensive, these breeds are more flavorful and worth the investment.

Chapter 7

RAISING TURKEYS

Now that you have decided to raise turkeys, you will need to do some preparation prior to their arrival on the farm. You will first need to determine if you can raise these large birds on your property. As outlined in Chapter 1, you will need to do the same footwork of tracking down the government office in your locality to ask if you can build a turkey house and keep turkeys on your property. Turkeys are large birds and will need plenty of pen space and a lot of space to roam to get the proper amount of exercise to remain healthy. The same equipment used to raise chickens can be used to start a flock of turkeys. *You can find a list in Chapter 1 and adapt it to your turkey flock.*

One important note: Turkeys should not be raised with chickens, as chickens carry the organism responsible for the disease "blackhead," which is further discussed in this chapter. Blackhead does not affect chickens, but it is a highly fatal disease in turkeys.

The Turkey House

Newly hatched poults up to 8 weeks of age will need to have a building to live in. You can use existing buildings on your property or buy or purchase a coop. A chicken coop, as described in Chapter 3, will be fine for young turkeys, but they will quickly outgrow these types of coops. A corner of a large barn, a feed room, or even a garage can be used to brood poults. Remember to remove all power equipment, cars, tractors, or ATVs that can cause fumes to collect in the building.

The top priorities of a turkey house are that it can be easily cleaned and disinfected, it has adequate ventilation and light, and it is predator-proof. The floor should be constructed of waterproof wood or concrete. The walls and roof should be of solid construction and weatherproof. Doors and windows should be well constructed and have a lock. Predator-proofing consists of snug doors and windows. Any cracks along the foundation should be sealed with cement or securely nailed boards. Even small openings of 1 inch can let in rats and weasels that will find their way in to eat feed and kill poults.

After eight weeks, your poults will outgrow their brooding pen. You can either enlarge the existing pen area or use a sun porch for your turkeys, which is a raised chicken-wire cage attached to a building or a three-sided shed. This is especially useful if you plan to raise your adult turkeys on pasture, as the sun porch will acclimate the young turkeys to the weather. Large, commercial turkeys should not be raised on chicken-wire floors because they will tend to develop breast blisters and foot and leg problems. Instead, they should be placed in a modified sun porch with a concrete pad for a floor.

The minimal space requirement for a turkey sun porch is 5 square feet of floor space per bird and 3 ½ feet of space above their heads.

The floor is constructed of sturdy wire or wooden slats to allow manure to drop to the ground. The floor should be raised 3 feet above the ground so droppings can be easily removed with a rake or shovel. The feeding and watering troughs should be placed into the sides of the sun porch, so they can be easily accessed from the outside. It should be located in a sunny area and ¼ of the porch should be covered to allow the turkeys to access shade if the porch becomes too warm. The maximum number of turkeys in one sun porch should be kept to less than 100 to prevent overcrowding and competition for space and food. Roosts of the same type used for chickens should be placed inside the building to allow the turkeys to roost at night. During rainy weather, poults under 12 weeks old should be locked inside the pen to prevent them from become chilled.

Portable coops

A portable or mobile coop is a convenient way to house your turkeys to allow them access to pasture or grass, much like it is for chickens. This is also a great way to distribute manure and to keep the coop clean. One such mobile turkey-housing unit is called a hoop house. It is a structure made of three hoops of 1-inch electrical conduit attached to 6-inch high baseboards to form a 10- by 12-foot building. The hoops are then covered with chicken-wire fastened with plastic ties. Then the entire structure is covered with a tarp to protect the turkeys from the elements. The open ends are enclosed with chicken wire. One end should have a wood frame for a door built into the end. This door should be able to be securely locked at night. Roosts should also be placed inside the building to give the turkeys a place to roost at night. Once the turkeys have come into the pen, lock them inside to keep predators away during the night.

This type of building makes a lightweight, highly portable pen. It should be staked down to prevent winds from blowing it over. Feeders should be placed inside the building to encourage the turkeys to enter the shelter. Place the feeders away from the roosts to prevent contamination with manure. Move the pen every two to three days to a fresh patch of pasture. Turkeys are birds of habit, so do not move the shelter too far, or they may refuse to return to the pen at night.

Many chicken coops could also be used for turkeys, granted they provide a large enough space. *Refer back to Chapter 3 for more ideas of poultry coops.*

Preparing for poults

Prior to purchasing poults (baby turkeys), you need to determine if you have the room to raise turkeys. Unlike chickens that need a minimal amount of space, turkeys are quite large birds. If you plan to keep a few breeds, they will need some room to roam. Allow 20 square feet of pen space per adult bird for indoor housing. Otherwise, fighting may develop among the birds, and it will be hard to keep the pen sufficiently clean of droppings.

Turkey poults will need to be confined and carefully monitored to decrease the death rate during their early life. Like other domesticated poultry, the young can suffer from cold weather, disease, and lack of water or food intake. The first few weeks until they are fully-feathered out are the most critical in the life of poults.

A young turkey's primary needs are to stay warm, dry, fed, and watered. You can keep them dry by using a layer of wood shavings, chopped straw, or sawdust on the floor for bedding. Turkeys are curious and might peck at the bedding (or litter) instead of their feed the first few days. Some producers will cover the bedding with cloth or paper to keep the poults from eating the litter. This can be removed in a few days when you are certain all the young turkeys are eating and will be more interested in eating feed than tasting the bedding.

To keep them warm, a corral made from 18-inch-high cardboard or straw bales can be used to confine the poults in the feeding and drinking area for the first few days. It is also a useful strategy to keep the birds near the heat source. A heat lamp with a 250-watt light bulb will be sufficient. It should hang so the bottom of the bulb is about 1 foot above the young bird's heads, adjusting it as needed if the birds huddle under the light by lowering the heat lamp, or if they stay away from the bulb by raising the heat lamp.

To keep them fed and watered, place plenty of feeders and waterers down for the new poults — two each for every ten poults will be sufficient. Observe the birds closely the first day or two to make sure they find the food and water. Place handfuls of feed on pieces of cardboard near the feeders to make it easier for the poults to find the feeder. Feed a commercial starter to the birds to ensure they get all the nutrients they need. You can purchase turkey starter mix from a feed store or grain elevator. After the birds are eating well, the feeders should be placed on bricks or hung to keep the birds from spilling the feed onto the floor. Raise the feeders enough so that the bird's heads are higher than their rears when feeding. One-gallon waterers are good for starting the young poults, but you will soon find you will need larger waterers to satisfy these fast-growing, large birds' thirst.

Behavior and breeding

Turkeys are naturally inquisitive and friendly birds, but they need to be trained to respect humans; when mature, they can be a bit territorial and terrorize pets and small children. Grabbing the turkey and pinning it to the ground can accomplish this, but be careful not to let the turkey hurt you with his spurs in the process. Be careful of their flapping wings as well, as these strong appendages can deliver quite a wallop to your face.

You may be surprised to learn that a turkey has a broad variety of vocalizations aside from the familiar gobble sound. The turkey also purrs, yelps, whines, cackles, tut-tuts, and clucks. Males have an air sac inside the chest with which they can make a drumming sound through air movement. Males will gobble often in the springtime during the breeding season. The gobble noise can carry for up to a mile. Females seldom gobble; instead, they tend to yelp.

During the mating season, the male turkey will strut with his chest puffed up and tail fanned. As a result of mating, the hen will suffer scratches, bruises, or defeathering from the claws of the male turkey. Many producers will fit their turkey hens with a saddle to prevent injuries, which is a leather or canvas jacket that slips under the wings and over the back. If you use a saddle, be sure to examine your hens frequently underneath and around the saddle to make sure it is not rubbing the skin or damaging the feathers. Turkey saddles can be ordered from Stromberg's (**www.strombergschickens.com/stock/turkeys.php**).

Artificial insemination is performed by collecting semen from the tom by massaging the turkey's abdomen and back over the testes with the hands. This causes the copulatory organ to protrude where the semen is "milked" from the organ by gentle hand pressure. The semen is collected with an aspirator or in a small container. To inseminate the female, hand pressure is applied to the left side of the abdomen around the vent or anus. This causes the cloaca – the common chamber of the turkey's digestive, urinary, and reproductive systems – to push out of place, and the oviduct – the long tube where egg formation occurs – protrudes. A syringe or plastic straw is inserted about 1 inch into the oviduct, and the semen is pressed out of the syringe or straw. As the semen is released, the pressure around the vent is released. The hens are inseminated at regular intervals for ten to 14 days to achieve optimal egg fertility.

Nutritional Requirements

Feeding turkeys the proper feed and in the right amounts will allow them to develop properly and prevent many diseases. Feed accounts for 70 percent of the cost of raising and keeping turkeys but *never* skimp on feed for these large birds. Large adult male turkeys can eat up to 1 pound of food a day while small adult hens will consume about ⅓ of a pound of feed. There are different feeding strategies depending on if you are starting poults, raising turkeys for meat production, or raising breeding turkeys. Regardless of stage of growth or your purpose for raising turkeys, there are some essential nutrients every turkey needs for life, growth, reproduction, and production. A lack of any of these nutrients will negatively affect the turkey's growth, reproduction, and health. These nutrients are water, protein, carbohydrates, fats, minerals, and vitamins. *Refer to Chapter 3 for more details on these nutrients.*

Feed companies can be valuable sources of information on how to feed turkeys. The starter and growing feed for turkeys should contain a coccidiostat to help control coccidiosis, a parasite that causes growth retardation, diarrhea, and death loss. Some feed companies that provide turkey rations include:

- Purina Mills: **www.purina-mills.com**
- Nutrena® Animal Feeds: **www.nutrenaworld.com/nutrena**

Turkey rations

Starter ration

Newly-hatched poults need to be taught to eat. A commercial turkey-starter with 28% percent protein is highly advisable to make sure the poults get the right nutrients in the proper amounts. The feed should be a mash type as opposed to a pellet version. If the baby turkeys do not eat their dry food eagerly it can be moistened with milk. Finely chopped dandelion leaves or green alfalfa can be sprinkled on top of the feed during the first few weeks. Only feed the poults as many greens as they will eat within an hour. In addition, provide the poults with a separate feeder with grit (to help the gizzard grind their food) and crushed oyster shells starting on day three of placing the poults in their brooding pen.

Grower ration

Starting at 2 months of age the turkeys can be gradually switched over to a growing ration. The feed should have a 24 percent protein level and can be a pellet version. If possible, they should also be allowed access to pasture at this time to allow them to forage for green growing grass or alfalfa. Cracked corn can also be offered in a separate feeder. Continue to provide grit and oyster shells as well.

Breeding ration

Adult turkeys (around 6 months of age) intended for breeding should be fed a turkey ration with 16 to 17 percent protein. Many times the same feed will be acceptable for all breeding poultry species. Supplement this ration with small amounts of alfalfa – hay or fresh – and continue to provide grit and oyster shells.

Confinement versus Pasture/Free Range

Depending upon your available building and pasture land, you may choose to raise your turkeys strictly inside a building, partially in a building and partially outside, or strictly on pasture (when they are mature enough). Each type of housing has its own good points and bad points. Only you can decide which type will fit into your plans for your turkeys.

Turkeys can be raised in strict confinement, using barns to house the birds. The most important environmental aspect when raising turkeys strictly inside is to give the barn adequate ventilation and adequate floor space. Adult turkeys will need 2 to 3 square feet of floor space per bird. To adequately ventilate the barn, you may need to install one or two exhaust fans, unless you have a barn with sidewall curtains that can be rolled up. Remember: If it is uncomfortable for you to breathe in the barn after you have been in there for about 30 minutes, it will affect your turkeys' health.

The bedding used for confinement housing should be kept in good condition. Straw, hay, or pine shavings all make acceptable bedding material. Dry and clean bedding will prevent common problems associated with confinement rearing such as breast blisters, skin blemishes, and soiled feathers. It will also help prevent respiratory disease from developing from ammonia fumes. Roosts are not necessary in confinement housing because the birds will be safe from predators and will not have to develop roosting behavior.

If you have the pasture space, turkeys, particularly the heritage breeds, are amenable to being raised outside. As a start, plan on stocking 100 birds per 1 acre. The best foraging plants for a turkey pasture are legumes like alfalfa or clovers and grasses like Timothy grass, orchard grass, or rye grass. The area where you plan to raise your turkeys should be fenced in using woven wire with small openings — make sure a young turkey cannot escape. A shelter, large enough to house all the turkeys at once without being crowded, should also be provided in case of poor weather. A portable coop will work well to shelter your turkeys on pasture and allow you to rotate pasture as it is grazed down.

Feeders should be provided and scattered throughout the field. Scattering feeders prevents turkeys from crowding near them and destroying the pasture around them. They should also be covered to prevent rain from damaging the feed and raised from the ground so they are level with the turkey's backs. Allow 6 inches of feeder space per bird. Trough-type feeders are good for the range since turkeys can access each side. Even if you only have a few birds, have at least two feeders available to prevent one turkey from bullying the weaker ones away from the feed.

Waterers and feeders should also be scattered throughout the pasture area, rotating their position frequently to prevent muddy areas around them. Waterers should be cleaned frequently with a bleach solution or a disinfectant. Fresh water needs to be available at all times. Five adult turkeys will drink 1 gallon of water daily, so plan the number of waterers accordingly. During warmer weather, the turkeys will drink more, so check the water supply twice daily to make sure they have not run out.

Pastures should be rotated every ten days to prevent a buildup of manure, mud, and disease-causing organisms. Each pasture should remain fallow (unused) for a month between uses. This will allow time for grasses to regrow and will let the sun kill off any disease-causing microorganisms.

Breeding Turkeys and Incubating Eggs

You may plan on breeding a few of your turkeys to replenish or grow your flock. Turkeys are harder to breed than chickens so you will want to carefully study the guidelines to turkey breeding before embarking on a breeding program with your flock.

About six months prior to your desired incubation start date, you will want to select your turkey breeds. Turkeys normally begin laying eggs during March and April. You will need one tom to breed ten hens. If you have the heavy commercial strains of turkeys, such as the broad-breasted bronze or the broad-breasted white, they will not be able to breed naturally. Instead, these breeds need to be artificially inseminated, as described earlier in this chapter.

The turkeys you choose for breeding should be healthy and free from any leg, feather, or head deformities. They should have a well-filled-out breast and straight backs and legs. They should be able to walk normally. The turkeys should also be true to their breed standards in size, weight, and feather coloring. Once you have chosen your breeders they should be separated from the rest of the flock six weeks before you expect to begin collecting eggs. Place them in a clean and disinfected pen that has 3 to 4 inches of bedding. The recommended minimal space requirement for a breeding turkey is 6 to 8 square feet. They can have access to an outside pen enclosed with chicken wire, but they need to be locked inside the pen at night. Each bird should have 4 to 5 square feet of yard space. Begin feeding them a breeder ration at this time as well.

A turkey tom strutting in a field.

Turkeys, both males and females, need a certain amount of light each day to stimulate their reproductive tracts; 14 hours of light a day, either natural, artificial, or a combination of both, is needed. A 50-watt light bulb (one per 100 square feet) will provide enough artificial light for the birds. Toms will need this lighting five weeks prior to breeding to ensure sperm production is at its peak for breeding. Hens need to be exposed to this lighting regime three weeks prior to breeding. You may need to place the tom turkey in a separate pen to ensure he gets the longer light period for the entire five weeks, but reintroduce him to the hens when you begin providing them with this light.

You will need to place nest boxes inside the pen for the turkey hens. These nests do not need to be covered, but hens do prefer to lay the eggs in a dark area. A simple 2- by 2-foot box made of wood can be constructed. Provide one nest per five hens and place them in a secluded part of the pen away from roosts and feeding/watering areas. Place them in the pen one month prior to egg production so the turkey hens will get used to them. The nest boxes should be filled with a 2- to 3-inch layer of bedding – pine shavings are the preferred bedding type for the nest boxes. Make sure to maintain this bedding in excellent condition by promptly removing soiled or wet material. This will assure that the eggs are kept as clean as possible. To make it easier on you when collecting eggs, place the nest boxes in an accessible area that does not require you to enter the pen.

Most heritage breeds of turkey hens will lay from 70 to 100 eggs during the five month egg-laying period. The broad-breasted bronze and the broad-breasted white turkey hens will lay around 90 to 100 eggs. Eggs should be collected twice daily from the nesting boxes to minimize breakage and soiling. This will also help decrease broodiness in the turkey hens. A broody turkey hen will stop laying eggs in preference of hatching the eggs she has already produced.

As discussed in the turkey behavior section, turkey saddles protect the hen turkey from becoming injured during mating. Clipping the tom turkey's claw and blunting his spurs by filing them will also help minimize any scratches or injury to the hen. If you find any scratches or injury while inspecting the hen, immediately treat the injury with an antiseptic spray. A hen can be used to incubate the eggs, but many turkey hens do not properly tend to their eggs. If a hen does become broody and refuses to leave the nest you will need to remove her from the breeding pen and put her in another pen near the breeding pen. Keep her there without food and little water until this ends, which is usually in five to seven days. Then she can be returned to the breeding pen to resume mating and laying eggs.

After you collect the eggs, inspect them for any cracks, surface damage, or abnormal shape. If you find any abnormalities, discard the eggs. Clean off any manure on the eggshell with a washcloth or rag dipped in lukewarm water. Place the eggs in a jumbo egg cart with the pointed ends down. These can be stored in a cool part of the house for up to a week before placing them in the incubator.

Incubation of eggs

Whether you have collected eggs from your breeding turkeys or you have purchased fertilized eggs from a hatchery, a fun project is to incubate eggs. For supplies you will need an incubator, a hygrometer for measuring humidity inside the incubator, and a regular thermometer. It takes 25 to 28 days for an egg to hatch a poult. You will need to carefully monitor the incubator during these four weeks to achieve the best hatch.

As with incubating chickens, allow the incubator to warm up for four days prior to placing the eggs inside. Read the instructions that come with the incubator carefully, as each manufacturer will have specific requirements on their product. Incubators generally come in two types: a still-air incubator or a fan-forced incubator. A fan-forced incubator will cost more, but for small batches of eggs a still-air incubator will give an adequate hatch. Maintain the temperature in the incubator at 99 to 99.5 degrees Fahrenheit.

It cannot be stressed enough to carefully monitor the temperature and humidity during the incubation period. Poults will not hatch or will be unhealthy in an improperly maintained incubator. The humidity level inside the incubator should be maintained at 50 to 55 percent.

Turn the eggs twice a day to prevent the poults from sticking to the side of the egg. Only open the incubator door during these time periods. Opening the door allows the essential heat and humidity to escape. It will take a few hours for the incubator to return to the proper levels of heat and humidity. Too many of these temperature and humidity swings may cause problems with the hatch. After one week of incubation, check the eggs with candling to see if they are fertilized. *See Chapter 4 for a description of candling.*

Three days prior to expected hatch date, stop turning the egg to allow time for the poults to get into hatching position. When the poults start to break their shells, do not attempt to help them out of the shell. If the crack a poult is making is toward the floor, gently reposition the egg so the crack faces up; then leave them alone. They need to break out of the egg on their own to develop the strength they need to survive in their new world. If a poult cannot come out of its shell on its own, it will probably not be able to survive.

The hatching process takes about five to ten hours. Leave unhatched eggs in the incubator for two to three days after the first poult hatches. After that, remove the eggs and discard them – these poults have died inside the cell. If your poults hatch wet and mushy, the humidity inside the incubator was too high. Remove the moisture source from the incubator if the poults fail to fluff up their down. Once their down is fluffed, and when they are running around the incubator, transfer them to the prepared brooder area. The poults should be ready for transfer a few hours after hatching.

LET'S TALK TURKEY—A CONSUMER GUIDE TO SAFELY ROASTING A TURKEY

FRESH OR FROZEN?

Fresh Turkeys

- Allow 1 pound of turkey per person.
- Buy your turkey only one to two days before you plan to cook it.
- Keep it stored in the refrigerator until you're ready to cook it. Place it on a tray or in a pan to catch any juices that may leak.
- **Do not buy fresh pre-stuffed turkeys.** If not handled properly, any harmful bacteria that may be in the stuffing can multiply very quickly.

Frozen Turkeys

- Allow 1 pound of turkey per person.
- Keep frozen until you're ready to thaw it.
- Turkeys can be kept frozen in the freezer indefinitely; however, cook within one year for best quality.
- See "Thawing Your Turkey" for thawing instructions.

Frozen Pre-Stuffed Turkeys

USDA recommends only buying frozen pre-stuffed turkeys that display the USDA or State mark of inspection on the packaging. These turkeys are safe because they have been processed under controlled conditions.

DO NOT THAW before cooking. Cook from the frozen state. Follow package directions for proper handling and cooking.

Allow 1¼ pounds of turkey per person.

Thawing Your Turkey

There are three ways to thaw your turkey safely — in the refrigerator, in cold water, or in the microwave oven.

In the Refrigerator (40°F or below) Allow approximately 24 hours for every 4 to 5 pounds	
4 to 12 pounds	1 to 3 days
12 to 16 pounds	3 to 4 days
16 to 20 pounds	4 to 5 days
20 to 24 pounds	5 to 6 days

Keep the turkey in its original wrapper. Place it on a tray or in a pan to catch any juices that may leak. A thawed turkey can remain in the refrigerator for one to two days. If necessary, a turkey that has been properly thawed in the refrigerator may be refrozen.

In Cold Water Allow approximately 30 minutes per pound	
4 to 12 pounds	2 to 6 hours
12 to 16 pounds	6 to 8 hours
16 to 20 pounds	8 to 10 hours
20 to 24 pounds	10 to 12 hours

Wrap your turkey securely, making sure the water is not able to leak through the wrapping. Submerge your wrapped turkey in cold tap water. Change the water every 30 minutes. Cook the turkey immediately after it is thawed. Do not refreeze.

In the Microwave Oven

- Check your owner's manual for the size turkey that will fit in your microwave oven, the minutes per pound, and power level to use for thawing.
- Remove all outside wrapping.
- Place on a microwave-safe dish to catch any juices that may leak.
- Cook your turkey immediately. Do not refreeze or refrigerate your turkey after thawing in the microwave oven.

REMINDER: Remove the giblets from the turkey cavities after thawing. Cook separately.

Roasting Your Turkey

- Set your oven temperature no lower than 325°F.
- Place your turkey or turkey breast on a rack in a shallow roasting pan.
- For optimum safety, stuffing a turkey is not recommended. For more even cooking, it is recommended you cook your stuffing outside the bird in a casserole. Use a food thermometer to check the internal temperature of the stuffing. The stuffing must reach a safe minimum internal temperature of 165°F.
- If you choose to stuff your turkey, the ingredients can be prepared ahead of time; however, keep wet and dry ingredients separate. Chill all of the wet ingredients (butter/margarine, cooked celery and onions, broth, etc.). Mix wet and dry ingredients just before filling the turkey cavities. Fill the cavities loosely. Cook the turkey immediately. Use a food thermometer to make sure the center of the stuffing reaches a safe minimum internal temperature of 165°F.
- A whole turkey is safe when cooked to a minimum internal temperature of 165°F as measured with a food thermometer. Check the internal temperature in the innermost part of the thigh and wing and the thickest part of the breast. For reasons of personal preference, consumers may choose to cook turkey to higher temperatures.
- If your turkey has a "pop-up" temperature indicator, it is recommended that you also check the internal temperature of the turkey in the innermost part of the thigh and wing and the thickest part of the breast with a food thermometer. The minimum internal temperature should reach 165°F for safety.

- For quality, let the turkey stand for 20 minutes before carving to allow juices to set. The turkey will carve more easily.
- Remove all stuffing from the turkey cavities.

Timetables for Turkey Roasting

(325°F oven temperature)

Use the timetables below to determine how long to cook your turkey. These times are approximate. Always use a food thermometer to check the internal temperature of your turkey and stuffing.

Unstuffed	
4 to 8 pounds (breast)	1½ to 3¼ hours
8 to 12 pounds	2¾ to 3 hours
12 to 14 pounds	3 to 3¾ hours
14 to 18 pounds	3¾ to 4¼ hours
18 to 20 pounds	4¼ to 4½ hours
20 to 24 pounds	4½ to 5 hours

Stuffed	
4 to 6 pounds (breast)	Not usually applicable
6 to 8 pounds (breast)	2½ to 3½ hours
8 to 12 pounds	3 to 3½ hours
12 to 14 pounds	3½ to 4 hours
14 to 18 pounds	4 to 4¼ hours
18 to 20 pounds	4¼ to 4¾ hours
20 to 24 pounds	4¾ to 5¼ hours

It is safe to cook a turkey from the frozen state. The cooking time will take at least **50 percent longer** than recommended for a fully thawed turkey. Remember to remove the giblet packages during the cooking time. Remove carefully with tongs or a fork.

Optional Cooking Hints

- Tuck wing tips under the shoulders of the bird for more even cooking. This is referred to as "akimbo."

- Add ½ cup of water to the bottom of the pan.
- If your roasting pan does not have a lid, you may place a tent of heavy-duty aluminum foil over the turkey for the first 1 to 1½ hours. This allows for maximum heat circulation, keeps the turkey moist, and reduces oven splatter. To prevent overbrowning, foil may also be placed over the turkey after it reaches the desired color.
- If using an oven-proof food thermometer, place it in the turkey at the start of the cooking cycle. It will allow you to check the internal temperature of the turkey while it is cooking. For turkey breasts, place the thermometer in the thickest part. For whole turkeys, place in the thickest part of the inner thigh. Once the thigh has reached 165°F, check the wing and the thickest part of the breast to ensure the turkey has reached a safe minimum internal temperature of 165°F throughout the product.
- If using an oven cooking bag, follow the manufacturer's guidelines on the package.

REMEMBER! Always wash hands, utensils, the sink, and anything else that comes in contact with raw turkey and its juices with soap and water.

For information on other methods for cooking a turkey, call the USDA Meat and Poultry Hotline
1-888-MPHotline (1-888-674-6854)
TTY: 1-800-256-7072
www.fsis.usda.gov

Storing Your Leftovers

- Discard any turkey, stuffing, and gravy left out at room temperature longer than two hours; one hour in temperatures above 90°F.
- Divide leftovers into smaller portions. Refrigerate or freeze in covered shallow containers for quicker cooling.

- Use refrigerated turkey and stuffing within three to four days. Use gravy within one to two days.
- If freezing leftovers, use within two to six months for best quality.

Reheating Your Turkey

Cooked turkey may be eaten cold or reheated.

In the Oven

- Set the oven temperature no lower than 325°F.
- Reheat turkey to an internal temperature of 165°F. Use a food thermometer to check the internal temperature.
- To keep the turkey moist, add a little broth or water and cover.

In the Microwave Oven

- Cover your food and rotate it for even heating. Allow standing time.
- Check the internal temperature of your food with a food thermometer to make sure it reaches 165°F.
- Consult your microwave oven owner's manual for recommended times and power levels.

Are you ready?

If you have read through the information on turkeys and are ready to start, then go for it. Like most novice poultry raisers, you will learn best by doing — and talking to other turkey growers. You may find starting out with a small flock of less than ten birds will allow you to experience raising turkeys and let you determine if turkeys are for you. If you are still undecided if turkeys are for you, you might want to consider raising another type of poultry mentioned in this book.

Part 3

WATERFOWL

Ducks and geese are quite a bit different from chickens and turkeys. They are all birds, but waterfowl have different requirements to be raised successfully. They are great birds to have around for meat, eggs, and pest control. Ducks make great pets and enjoy eating weeds. Geese learn to guard their territory and can keep intruders off of your property. Raising either type of bird will positively impact your life. This section will give you all the information you will need to raise a duck or goose all the way from egg to adult.

Chapter 8
DUCKS

Ducks are closely related to swans and geese; in fact, biologists have had difficulty classifying the three species into different categories. The three species belong to the biological family Anatidae, meaning birds that swim, float on the surface of the water, and (in some birds) dive in water for food. For the most part, the birds in this group eat plants and grains and are monogamous (one mate) breeders under natural conditions.

It is no wonder biologists have trouble classifying the three species, as there are more than 100 species of wild ducks alone. Despite the large number of wild ducks, all domestic ducks, except for the Muscovy duck, were domesticated from the wild Mallard. The Mallard belongs to a group of wild ducks called dabblers. This describes their feeding habit of dabbing their bill in the water to filter food particles from the water. The Mallards also have an appetite for grasses, insects, bugs, worms, small fish, toads, snails, frogs, and even snakes.

The Chinese are credited with first domesticating wild Mallards around 1000 B.C., but ducks were also domesticated around the same time in the Middle East. The ducks were first domesticated for meat and eggs use, although later the ducks were used to control pests in rice paddies. Even today, people will herd flocks of ducks to the rice paddies. In the field the ducks feast on insects, snails, slugs, small reptiles, and waste rice, providing great pest control without the use of harsh chemicals.

The Muscovy duck has a far different history. The Muscovy is actually a perching duck and will roost and nest in trees. They are a native of Central and South America. It is thought they were domesticated in pre-Inca Peru to be used as pets. The Muscovy is a carnivorous duck, eating small mammals, snakes, frogs, flies, and other insects. They can grow to be larger than most breeds of duck. In the United States, many people keep this breed as pets because they do not make much noise and have distinct personalities.

This African white-faced duck is foraging in shallow water.

Why Raise Ducks?

A duck, when full grown, can weigh 4 to 11 pounds. They can live as long as 12 years and can provide a reliable source of eggs and meat. Some breeds have been developed to be prolific egg-layers, while other breeds have been developed to provide substantial quantities of meat. If you plan to collect eggs from ducks, the Indian runner and khaki Campbell are two breeds known for their prolific egg production. People can eat duck eggs, much like they eat chicken eggs, and the taste is similar. However, there will most likely not be a market for the eggs unless you develop one. Duck eggs have a higher yolk fat content and higher white protein content than chicken eggs. When cooked, the whites do not become as stiff as chicken eggs. If the duck has been eating a lot of algae, worms, or grubs, the eggs may have a slightly bizarre taste.

Duck meat is higher in iron, niacin, and selenium than many other types of meat. Duck meat is all dark meat and is richer than chicken or turkey due to the higher fat content, which gives the meat a distinctive taste popular in many gourmet restaurants. Ethnic markets are particularly interested in obtaining a steady supply of quality duck and goose meat. Good duck meat breeds include Pekin, Rouen, Muscovy, and Aylesbury. In addition to providing your family with meat and eggs, ducks provide excellent insect and small pest control. Adult ducks can derive a lot of their diet from eating pests while ridding your yard or garden of harmful critters. They are vulnerable to predators when on land so you will want to keep an eye on them when they are out on pest control.

Some prefer duck eggs over chicken eggs.

Duck Breeds

When deciding which breed of duck you would like to raise, there are two main types to consider. Ornamental ducks are kept for the pleasure of keeping waterfowl. Their striking plumage and interesting behavior are their primary benefit to their humans, and generally they do not make good meat or egg producers. Utility, or commercial ducks, have been bred for meat, down, or egg production. Ducks come in many different feather colors; males and females can be distinguished by this trait. Further classification includes:

- Bantam: less than 2½ pounds
- Light weight: 4 to 5 pounds
- Medium weight: 7 to 8 pounds
- Heavy weight: 9 pounds or more

Indian runner

Drake: Up to 5 pounds
Duck: Up to 4 pounds
Eggs: 200 per year

This breed of duck is unique in its upright stance, looking like a bowling pin on webbed feet. It was developed from the wild mallard in the East Indies two centuries ago. Because of their physical conformation, they run rather than waddle, which makes walking through fields an easy task for Indian runners. They are active foragers, especially of insects, snails, and slugs. This breed does not put on a good deal of flesh, considering they are a lighter duck. Indian Runners are great egg-layers – sometimes even outlaying chickens. However, they are not considered good egg-sitters, so eggs from a female Indian Runner may need to be incubated by another duck or in an incubator.

Khaki Campbell

Drake: Up to 5 pounds
Duck: Average 4½ pounds
Eggs: 300 per year

The khaki Campbell is a breed developed in England by Adele Campbell in the late 1800s. Campbell crossed her Indian Runner hen with a Rouen Drake in order to produce ducks that would lay well and have bigger, meatier bodies. Her breeding strategy worked because the resulting breed, the khaki Campbell, is an excellent layer, and the females will readily sit on the eggs, unlike the Indian runner duck. The khaki Campbell does have a flighty temperament and needs room to forage. They are also adaptable to variable climates, performing well in hot, dry deserts, wet, tropical environments, or in cold winter weather. As an added bonus to their egg production, the khaki Campbell is a prolific insect, slug, and algae eater.

Pekin

Drake: 12 pounds or more
Duck: 11 pounds or more
Eggs: 200 per year

The Pekin is the most common breed of domestic duck. These white ducks are great for meat production as they grow rapidly and pack on more meat per pound of feed than other ducks. The Pekin duck was developed in China from ducks residing in the canals of Nanjing. Farmers in China eventually domesticated the resulting duck. They are not as broody as other ducks, so they may not sit on a nest.

Rouen

Drake: 6 to 10 pounds
Duck: 5 to 8 pounds
Eggs: 70 per year

The Rouen breed is similar in coloring to the wild Mallard. They were originally developed in France and were imported to England, where the breed was bred into the modern-day Rouen. There are two types of Rouen: the production and the standard. The production Rouen weighs between 6 and 8 pounds; the standard Rouen is much larger, weighing between 8 and 10 pounds. They are good meat producers but take from six to eight months to mature. This slow maturation rate has lead commercial duck growers to be reluctant to raise Rouens on a large scale for the meat market. The meat from the Rouen is leaner than other European breeds, making it a popular duck for restaurants.

Muscovy

Drake: 7 pounds or more
Duck: 3 pounds or more
Eggs: 80 per year

The Muscovy is unique in that it was not developed from Mallards. The breed is from South America, though its name indicates it is from Moscow. The breed has many names in different areas of the world, including the Barbary, the Brazilian duck, or the Turkish duck. The breed comes in a variety of colors, including black, white, chocolate, and blue, but they all have a distinctive red tissue above the beak and around the eyes. They do not swim much, as they have underdeveloped oil glands, which make their feathers less water-resistant than other breeds. They do have sharp claws, which they use to roost in tree branches. The females become broody an average of three times a year. They will incubate and hatch the eggs of other ducks or poultry species. Overall, the breed is easy to care for because they can find their own food, reproduce easily, and will care for their young.

Cayuga

Drake: 8 pounds
Duck: 6 to 7 pounds
Eggs: 120 to 150 per year

This breed was developed in New York in the 1800s from native ducks. They are considered to be a medium-weight duck primarily used as a meat bird. They have unique coloring with a greenish-blue sheen over dark black feathers. As they age, the feathers of the Cayuga tend to get lighter. Eggs from the Cayuga can be variable colors depending on the season. When they first start laying eggs, the eggs may have a gray or black color. As the laying season progresses, the eggs will start to lose this dark coloring and may even become white. Hens will become broody if they are allowed to sit on their eggs. This breed likes to eat grasses and bugs like worms and slugs.

Ancona

Drake: 6 pounds average
Duck: 5 pounds average
Eggs: 210 to 280 eggs per year

The Ancona has coloring similar to a Holstein cow: white feathers with spots. The spots can be black, blue, chocolate, lavender, or a combination of spots of two different colors. Even their bills can have black spots. It is a medium-weight duck with yellow legs. The Ancona is an excellent forager, and it is easily trained to stay close to home. Like the Indian Runner, it is an excellent egg-layer, and it produces a nice carcass for meat purposes. There is a need for conservation breeders to preserve this breed.

Appleyard

Drake: Up to 8 pounds
Duck: Up to 6 pounds
Eggs: 220 to 265 eggs per year

In the 1930s, Reginald Appleyard developed the Appleyard duck in England. It is a heavy duck that grows fast and produces an excellent carcass. Its coloring is similar to the Mallard, but it has a lighter silvery tint to the feathers, some call the breed the silver Appleyard. The Appleyard is an excellent forager and lays white-shelled eggs. The ducks will stay close to home as long as they have enough food.

Call

Drake: Up to 2 pounds
Duck: Up to 1½ pounds
Eggs: Maximum 80 eggs per year

Call ducks are bantam ducks. They have a small beak and a rounded head. The call duck almost looks like a child's stuffed toy duck and are primarily used as show or pet ducks. Due to their size, call hens do not lay many eggs per year – some may lay less than 20 per year. The number of eggs a hen will lay is widely varied. The call duck is easily tamed and will "chat" with their owners. They are extremely noisy (hence the name) and have loud high-pitched voices. The call duck comes in many varieties and the most common are white and grey. They were developed in Holland to be used as decoy ducks. Their high-pitched voices carry over long distances. Hunters used the call duck to "call" wild ducks to the hunting site.

Crested

Drake: 7 pounds average
Duck: 6 pounds average
Eggs: 100 or more per year

This medium weight duck has a puff of feathers at the back of the head. This puff, or crest, is a mutation that occurs in Mallards occasionally, but a crest on their head defines this breed. Embryos that carry a copy of the mutation from each parent die in the shell case. Because the crest is fatty tissue emerging from a gap in the skull, the mutation may also cause neck deformities and problems with balance. The crested is primarily kept as a pet. There are two recognized colors: black and white. It can be hard for beginners to breed the crested duck because an understanding of genetic information is required.

East Indies

Male bantam: 1 to 2 pounds
Female bantam: 1 to 1½ pounds
Eggs: 20 per year

Despite its name, the East Indies duck was developed in North America in the 1800s. It is classified as a bantam duck. It has a striking metallic green color with blue and black highlights. As hens age, feathers can turn white. The East Indies duck makes a good backyard pet although they are also good fliers so their wings will need to be clipped. This breed often is compared with call ducks because of their size similarities. They are not vocal like call ducks and they tend to be more docile than calls.

Golden cascade

Drake: 7 to 8 pounds
Duck: 5 to 6 pounds
Eggs: 200 eggs per year

This medium-sized duck is a fairly new breed developed in 1979. It was developed from crosses of Pekin, crested, khaki Campbell, and Rouen. They produce a nice carcass and are prolific egg-layers. At birth the ducklings are golden colored. Males have reddish-brown necks and backs and white chests. Females have fawn-buff necks and backs and cream colored chests.

Magpie

Drake: 6 pounds average
Duck: 5 pounds average
Eggs: 290 per year

This duck breed was developed in Wales and imported to the United States in the 1960s. It is a lightweight duck and is a prolific egg-layer. The magpie does make a good meat duck, but being a small duck there will be a smaller yield of meat. Their feathers are white with splotches of color; black is the most common color, but there are ducks with blue, silver, or chocolate splotches.

Saxony

Drake: Up to 8 pounds
Duck: Up to 6 pounds
Eggs: About 200 per year

The Saxony was developed in Germany. It is a heavy breed of duck and produces a good carcass yield. The male Saxony will have a powder blue head and a white ring around the neck. The back and body ranges from silver to oatmeal color. The female Saxony is buff-colored with some blue and white feathers. The feet of this breed are usually orange.

Swedish

Drake: Up to 8 pounds
Duck: 6 to 7 pounds
Eggs: 100 to 150 per year

The Swedish is a blue-colored duck with a white bib. They were popular in Europe for centuries and imported to the United States in 1884. It is a medium-sized bird that produces a good carcass. It is a great forager and its dark color may make it less vulnerable to predators, which will have trouble seeing the bird in grass. They have a calm temperament and do make good pets if tamed as ducklings.

Duck conservation

With the loss of smaller farms, many duck breeds are in danger of extinction like other poultry species. The American Livestock Breed Conservancy (**www.albc-usa.org**) is working hard to protect these threatened breeds. Their mission as a nonprofit organization is to conserve historic breeds and the genetic diversity of livestock. Hatcheries are also important in the preservation of duck breeds. They are generally owned and staffed by knowledgeable poultry enthusiasts who are interested in preserving certain breeds. Many will maintain a number of breeds for you to choose from when you place your order. Here is a list of hatcheries that sell popular and rare breed ducklings:

- Cackle Hatchery: **www.cacklehatchery.com**
- Dunlap Hatchery: **www.dunlaphatchery.net**
- Holderread Waterfowl Farm & Preservation Center: **www.holderreadfarm.com**

- Murray McMurray Hatchery: **www.mcmurrayhatchery.com**
- Sand Hill Preservation Center: **www.sandhillpreservation.com**
- Stromberg's: **www.strombergschickens.com**

Raising Ducks

Ducks have an oil gland at the base of the tail. They will rub their chins and cheeks over the gland collecting the oil, which they will then rub onto their feathers. This oil makes their outer feathers waterproof. A mother will rub some of the oil from her oil gland onto the down of her young until they are able to perform this function on their own at around four to five weeks when they are fully feathered out. Ducks need to keep their feathers in tip-top condition to keep them dry and warm, so they spend a substantial amount of time preening.

Preparing for ducklings

Ducklings are purchased the same places as chickens: directly from the hatchery or through feed stores. They will come in a straight run (shippings of both sexes), or they can be sexed if you want to have more females or males. Generally a straight run will be a few cents cheaper per duckling to purchase than those segregated by sex. For the average small-scale farmer, a straight run will be satisfactory for ducks raised for meat. The males generally will be heavier than females, but female ducks are perfectly acceptable for meat. If you want to breed ducks, you may want them sexed so you can have a proper ratio of males to female. One drake will mate with five to six females.

When purchasing your ducklings directly from the hatchery, ask when they will be mailed. It is safe to ship newly hatched waterfowl as long as they are properly packaged in sturdy cardboard containers with plenty of air holes. They can go without food for a day or two while being shipped, as after hatching they retain part of the yolk from the egg in their body. This gives them a food source, but you will want to make sure someone is home to receive the ducklings when they arrive in the mail. Open the shipping container in the presence of the mail carrier to ensure you have received the number of birds ordered and to check for any dead stock. If the number is less or if there are any dead birds, the postal carrier can give you a claim check to submit to the hatchery.

If you choose to purchase your stock from a store, closely scrutinize the conditions of its pen. It should be dry, and the feed and water containers should be clean and full. The ducklings should be active with no noticeable discharge from eyes or nose. Take a peek under their tails. The vent (anus) should be clean with no buildup of fecal material.

Prior to ordering or heading to the feed store to purchase your new additions, the pen, feeders, and waterers should be set up and ready. The pen should be in a draft-free, fully enclosed building with good ventilation and lighting. A corner of a garage or barn will work well provided you keep running motors out of the area so the birds are not subjected to fumes. You should allow 6 inches of space per bird, and increase this to 1 foot per bird after they are two weeks old. At four weeks of age, the ducklings should have 1½ feet of space in the pen, and by six weeks they will need 2 feet of space. The pen floor should be covered with an absorbent litter. Four inches of wood shavings, peat moss, or chopped straw will be sufficient. The litter will need to be maintained to eliminate wet, dirty spots. Add fresh litter as needed to maintain 4 inches of bedding.

Provide heat and light through a heat lamp with a 250-watt bulb or use a hover-type brooder. A hover brooder uses propane heat coupled with a metal pan (or hover) to direct and retain the heat over the young. Plan to use one heat lamp per 25 ducklings. Hover brooders typically come with instructions for chicks. Since ducklings are larger than chicks, brood half as many ducklings as you would chicks. Ducklings are taller than chicks, so you may need to raise the height of the heat lamp a few inches. The heat-source temperature should be between 85 to 90 degrees Fahrenheit. This temperature can be reduced by 5 to 10 degrees each week until the temperature is around 70 degrees Fahrenheit. After the sixth week and if the weather is mild, the birds will be fully feathered and will no longer need supplemental heat.

The birds should be confined near the heat, feed, and water sources during the first two to three days after arrival. Observe the birds closely to determine if the heat source needs to be raised or lowered. As with chickens, if the birds avoid the heat source and are lurking at the edges of the pen, raise the heat source a few inches. If they huddle under the heat source, lower it a few inches.

As flock birds, you might notice your ducks following one another.

Waterers should be full when the birds arrive. There are many types of reasonably priced waterers available. Do not use an open pan for young ducklings; they should not get wet when they are in the down stage, lest they become chilled. A chilled baby bird can quickly become hypothermic and die. Instead, use a waterer with a base wide enough in which the birds can dip their heads and bills. Adding commercial electrolyte or vitamin powder to the water the first few days can give the young birds a healthy boost.

Do not let young ducks have access to swimming water or leave them outside in the rain. The feathers of young ducks are not fully developed to protect them from water, especially during the down stage. If they have been hatched out by a mother duck, they can have access to swimming water with the adults because the mother will not let them remain in the water for too long, and she will protect them from rain. By 4 to 5 weeks of age, the ducklings will be feathered out and will be able to tolerate most weather conditions.

Nutritional Requirements

There are six classes of nutrients that are essential for life, growth, reproduction, and production. A lack of any of these nutrients will negatively affect the duck's growth, reproduction, and health. These are water, protein, carbohydrates, fats, minerals, and vitamins. *Refer to Chapter 3 for more details on these nutrients.*

Feed companies can be valuable sources of information on how to feed ducks and geese. Each company will have their own feeding strategies for each stage of a waterfowl's life, but generally they will provide starter, growing, finishing, maintenance, and breeding rations. Some feed companies that provide duck and goose rations include:

- Purina Mills: **www.purina-mills.com**
- Nutrena: **www.nutrenaworld.com/nutrena**

Duck rations

Starter ration

A good starter feed will set your young ducklings off to a good start in life. They are generally disease resistant so a medicated feed will not be necessary. In fact, certain medications found in chick starters can cause health problems in ducklings. In addition, young ducks need three times the amount of niacin as chicks. If you must feed a chicken starter ration, niacin will need to be added to prevent disease. This can be done through supplementing the feed with poultry vitamin mixes or adding 6 pounds of livestock brewer's yeast per 100 pounds of chicken feed. Niacin deficiency will cause the birds to have deformed, weak, and bowed legs. If not recognized or treated in time the ducks will remain permanently crippled.

During the first three weeks of life, the ducklings should be fed a feed ration of 20 to 22 percent protein. Keep feed in front of ducklings less than 2 weeks of age at all times. Avoid feeding them a finely ground mash as they may choke on it, and they will waste a lot of the feed. A coarse ground feed is best for the ducklings. Starting the first week of life, you can also feed small amounts of fresh growing grass or fresh clippings to the birds. Grit should also be kept in front of the ducklings at all times. This will assist their gizzards in grinding the feed.

Growing ration

After 4 weeks of age, the duckling's diet can be supplemented with cracked corn, and they can be switched to a grower ration. An ideal growing ration will have 16 to 18 percent protein. A plot of pasture enclosed by a 3-foot, woven-wire fence makes a great feed source for the ducks when they are about 6 weeks old. Ducks enjoy foraging and eating both bugs and plants. An acre of pasture can support up to 40 adult ducks. Good plants for pasture are brome grass, Timothy, orchard grass, bluegrass, and clover. Provide young growing ducks with a secure shelter at night or predators will decimate your flock.

Finishing ration

If your ducks are to be slaughtered for home use or for market they should be fed a finishing ration formulated for turkeys starting one month prior to slaughter. This will provide them extra nutrients allowing them to fatten prior to slaughter. Birds not intended for slaughter do not need to be fed a finishing ration and instead can be switched to a maintenance or a breeder ration. *To learn about butchering your birds, see Chapter 5.*

Maintenance and breeding rations

A maintenance ration — one intended to maintain ducks at current weight and nutritional level — for a duck consists of 13 to 14 percent protein. If ducks are allowed to forage for grasses and insects, a minimal amount of maintenance feed will need to be fed. For larger ducks, they require ⅓ pound of feed daily. Smaller ducks should be offered ¼ of a pound of maintenance feed a day. Ducks intended for breeding and egg laying need to be given a feed with 16 to 20 percent protein. This feed should be fed three weeks prior to breeding or egg laying. Large ducks will need ½ pound of this feed daily while smaller ducks will need ⅓ of a pound.

Duck Husbandry

Three ducks sitting in a row on a riverbank.

Ducks enjoy water not only for quenching their thirst but also for cleaning their bodies and exercise. Ducks cannot survive without access to bathing water. They will need a deeper trough to dip their bills in to drink than is required for chickens. It is their habit to splash water onto their heads and bills to clean themselves.

If you have a few birds (less than ten) you can use a hose and a small kiddie-pool to provide bathing water. This will need twice-daily cleaning and refilling. The ground surrounding the pool will quickly become muddy with the ducks hopping in and out of the pool. You can combat the mud issue by frequently relocating the pool to a new area or placing it on sand or gravel. Alternatively, a platform built of water resistant 2-inch by 4-inch boards and welded wire can be built. It should be 2 to 3 feet wider than the pool to allow a place for the ducks to step on as they enter and exit the pool. The birds should have a waterer for drinking. They should not be able to swim in this water but only submerge their head and drink water.

If you have more than ten birds, provide separate waterers and bathing tubs. The bathing tubs should be cleaned once a day. You can use a small pond for bathing, but the banks can quickly become damaged if too many ducks use the pond. To combat bank erosion, the bank can be reinforced with large stones. Ducks can be quite damaging to the shoreline as they dig in the mud in search of food.

Ducks are susceptible to predation, especially domestic ducks that have a limited ability to fly. A predator-proof pen should be provided for your ducks. They can roam during the day, but at night put them in a pen – nighttime is when predators such as dogs, weasels, coyotes, and raccoons are most active. Strong woven wire should be used with squares of less than ½-inch. Weasels are able to squeeze through any wire bigger than this size. The top of the pen should also be covered to prevent animals from climbing over the top.

Inspect the pen weekly. Check the pen or building for any signs of tunneling under the floor or foundation. Patch any holes with concrete, wood, or wire as soon as they are found. Make certain the doors and any windows are solid, that latches work, and that they are free from damage.

Also, you may prefer to keep your ducks penned and confined to a certain area during the daytime. Many types of fencing material can be used such as chicken wire, welded wire, woven wire, gamebird netting, and lightweight plastic fencing material. The fencing material should be a minimum of 2 feet in height to keep the ducks inside. If you need to keep dogs out, the fence should be 5 feet in height or higher.

The spacing between the wires should be ½ to 1 inch. Any bigger and the ducks can stick their heads between the wires. This may cause them to become trapped or even worse: A predator can bite the head of a duck that sticks its head outside the fence. The fencing material should be stretched taut and secured to posts (steel or wood) securely driven into the ground.

Keep in mind that ducks will need shade if temperatures approach 70 degrees Fahrenheit or above. Natural shade through trees or allowing access to a covered pen will be adequate. If the ducks are kept fenced in an opened area, you can build a simple shade for them using three 12-inch long 2-inch by 4-inch boards and a piece of ½-inch thick plywood. Farm Tek (**www.farmtek.com**), an agricultural supplier, also carries lightweight livestock canopies and covers composed of tough, weather-resistant plastic, which makes a simple and affordable shade cover for ducks.

If you are not raising your ducks for meat production or egg laying, adult ducks can forage for much of their food if they are allowed to roam. Feed should still be provided for the ducks, however, as there will be seasonal variations in food supply. Use a maintenance-type food as described in the nutrition section. To make feeding less complicated, a gravity feeder can be used. This type of feeder is a round-shaped container on a base that dispenses feed down as the ducks eat at the bottom.

Place the feeder(s) inside a pen to encourage your ducks to come to the pen in the evenings. It should be in a covered area so rain will not be able to destroy the feed. eNasco (**www.enasco.com**) carries many styles of feeders for poultry producers made from both galvanized steel and plastic. Feeders and waterers for adult ducks should be cleaned as they become soiled and disinfected once a month.

Breeding

On ducks intended for breeding, apply a wing or leg band so you can quickly identify them. Do this soon after hatching; the band used on each animal should have different numbers. Keep records regarding the parents of each duck, how many eggs each duck lays, if she is broody and for how long, and how many eggs are hatched for each female. Poorly performing ducks can be identified by their band and removed from the flock; this is also known as culling.

The tail features of ducks descended from Mallards will help you determine the sex of each bird. Drakes will have a few curled feather tips at the end of their tails while female's tails will lie flat. You can also sex a duck by examining the vent or anal opening. To sex a duck, hold the bird with the vent facing the person performing the sexing. Place the right thumb and first finger on either side of the vent and press firmly over the vent. The vent will then part slowly to expose the inner lining. Use the left thumb to gently pull back on the skin surrounding the vent. This will expose a pink-colored cloaca and the penis (a small protuberance) in males. Females have a genital eminence, or small fold of tissue. Adults are sexed in a similar manner but will struggle when caught. Note that geese can be sexed the same way.

Only those ducks in good physical shape should be kept for breeding. Legs should be straight and free of deformities, as should the beak and wings. They should comply with breed standards for coloring, body shape, and weight. A drake can breed five to ten females, the number varies based on the breed of duck. Ducks can be bred during their first year of life after they have reached maturity.

A mother duck with 9 ducklings.

Most duck eggs take 28 days to hatch, with the exception of Muscovy duck eggs, which take 35 days. If the female incubates her own eggs, make sure she has water and feed available near the nest. Pekin and Indian Runner ducks do not make good egg sitters, so you may need to have a foster mother incubate the eggs. Duck eggs can be brooded by broody chicken hens, but the eggs will need to be sprinkled with water every day as they have a higher humidity requirement while incubating than chickens.

If you plan to artificially incubate the eggs, the process is similar to chickens except for differences in humidity and temperature. Incubation requires 99.5 degrees Fahrenheit and 55 to 75 percent humidity. The eggs need to be turned four times a day. At day 25, the temperature should be lowered and the humidity slightly increased. Once the ducklings are hatched, allow them to dry in the incubator for one hour. Then they can be moved to their prepared brooding pen.

Removing Feathers

Saving the feathers from ducks and geese can be a useful idea. The process will occur after you kill the bird, when you are preparing to butcher it. *To learn more about butchering, see Chapter 5.*

The down from the breast area can be washed, dried, and saved for use in pillows or clothing. After butchering, geese and ducks can be dry-picked, but scalding the feathers first makes them release easier and cuts down on skin tears. If you do not want to save the down, the goose or duck can be waxed after the large feathers are removed. This process will remove the down and smaller feathers that are more difficult to remove.

Melted paraffin wax can be purchased and heated to 140 to 155 degrees Fahrenheit. The goose or duck should be dipped twice into the wax, then dipped into cold water to set the wax. When the wax sets to a flexible form, it can then be stripped off along with the down and feather. The wax can be strained of the feathers and down for reuse.

Are you ready for ducks?

If you have read through this chapter on ducks and are still revved up to add ducks to your barnyard, you are most likely ready to wade into duck farming. A duck adds a certain charm to your farm and is an excellent addition to any poultry venture. Ducks do not disappoint.

A flock of 11 female ducks swimming together.

Chapter 9
GEESE

Geese were this author's first foray into her own farm enterprise when she was 12 years old. Geese are hardy, lively birds that practically raise themselves after they lose their down. Not much maintenance work is required for the majority of a goose's life. They are also elegant birds and have a unique personality while being versatile, as you will soon find in the course of reading this chapter.

A family of Canada geese swimming through a pond.

History

The goose was domesticated both in eastern Asia and in northern Africa, Europe, and western Asia. In eastern Asia, the swan goose (*Anser cygnoides*) was domesticated to become the Chinese goose.

These geese possess a large knob at the base of their bill. The European-type goose was domesticated from the Greylag – or wild grey – goose (*Anser anser*) in northern Africa, Europe, and western Asia. Both types have been used since first domesticated for their meat, down, and eggs.

Archaeological evidence in Egypt has shown that geese were kept in ancient Egypt since 300 B.C. The Romans dedicated geese to the goddess Juno. Huge flocks of geese were raised in western Europe and slowly herded to Rome to supply this great city with meat and feathers. As time advanced, large flocks were also raised in southern England, Holland, and Germany and were driven to markets in large cities during the fall. Another important product of the goose was the quill, which was used for pens.

Despite being domesticated for centuries, the goose has not undergone the drastic changes seen in other domesticated livestock. The major changes include an increase in size, more fat deposition under the skin, selection for color (most notably white), and improved fertility. The domestic goose does tend to have a more upright posture than its wild brethren and generally is unable to fly.

In Asia, the goose is still an important livestock species. Markets do exist in the United States, particularly among immigrants from Asia and along the eastern seaboard. Geese often act as guards for property. For example, the company that brews Ballantine's Finest Blended Scotch Whisky has been using geese to guard their maturing products since 1959. The guards are nicknamed the Scotchwatch. When being sold for meat, geese are usually marketed for a fall market, particularly around Christmastime.

Domestic Goose Breeds

Emden

Gander: 30 pounds
Goose: 15 to 17 pounds
Eggs: 40 per year

The Emden was developed in Germany and Holland and is the most common commercial goose breed. They are typically white with orange bills and feet, blue eyes, and they grow rapidly, making them large meat-type geese. They make excellent barnyard alarms, as they can be protective of their territory and flock. They can be aggressive, especially a gander protecting his flock. Small children and pets should be watched when around a flock. Male Emden goslings have lighter gray down than the female goslings. It is also a good breed for a crossbreeding program because the breed matures early, is a good forager for food, and the females are good mothers.

Toulouse

Gander: 25 pounds
Goose: 20 pounds
Eggs: 35 per year

The Toulouse is noted for its cold tolerance and is a popular breed in the Midwest, although its origins are in France. The Toulouse breed has dark gray feathers on its back, lighter gray feathers on its breast, and white stomach feathers. It has a dewlap, a flap of skin, hanging under its lower jaw and a bulky body. The Toulouse was bred in France to produce foie gras. As such, it is not as good a forager as other geese breeds, but it does well when confined to a pen. The goslings also mature slower than other breeds. Toulouse geese can be clumsy and break eggs if the nest is not well padded.

Chinese

Gander: 12 pounds
Goose: 10 pounds
Eggs: 50 to 60 per year

This breed is the smallest of domestic geese. They have been called swan geese, as they carry their body upright (similar to swans). They are distinctive geese because they have a knob at the base of their beak. The knob on the male is larger than the females'. Chinese geese come in two colors: brown and white. The white variety has a more attractive carcass, as their pinfeathers are not as noticeable. Another added trait is they make excellent weeders and eat weeds from vegetable crops without causing much damage to the vegetables. The Chinese goose makes a good guard goose for the farmstead. They can easily fly over fences.

African

Gander: 22 to 24 pounds
Goose: 18 to 20 pounds
Eggs: 35 to 40 per year

The African goose is related to the Chinese goose, but it is a much larger breed. It has a distinctive knob on its forehead near the bill and a dewlap. It is about the same size and weight as the Emden breed. The colored variety of African has brown top feathers and a lighter underbelly. The white African variety has white feathers and an orange bill, knob, legs, and feet. Despite its name, it does not come from Africa, but its origin is murky. African geese can start breeding their first year and can produce eggs for many years. They are often raised for meat because their meat has a low fat content.

Pilgrim

Gander: 14 to 15 pounds
Goose: 10 to 13 pounds
Eggs: 35 to 40 per year

The pilgrim is the only breed of geese in which the male and female have different color feathers at maturity; adult males have mostly white feathers, while females have gray feathers. Being able to determine the gender of a bird based on the color of their feathers is called autosexing. Even when 1 day old, the breed can be sexed on color: Male goslings are gray-yellow with light bills, and female goslings are olive-gray with dark bills. It is also a calmer breed than most of the goose breeds, but will still sound an alarm at perceived signs of danger. They are good foragers and attentive mothers.

Sebastopol

Gander: 14 pounds
Goose: 11 pounds
Eggs: 40 per year

This unique breed has blue eyes and curly, twisted feathers that are usually white. Due to their unique feathers, they should have bathing water available to keep their feathers clean. Because of their feather pattern, they are more susceptible to chilling. Unlike other geese, water does not roll off their feathers. They are a gentle breed of goose and are not aggressive, which makes them more susceptible to predation. Females will brood the eggs of other geese. In fact, they may steal eggs from other nests and roll them into their own. Female goslings have darker down than the males.

Raising Geese

Raising geese can be a rewarding addition to your home or small farm. There are many similarities between raising geese and raising chickens and ducks, but these unique birds have some special needs that need to be taken into account.

These wild Greylag geese are enjoying water off a riverbank.

Handling aggressive ganders

Ganders raised and imprinted on humans can become aggressive toward humans because they view humans as rivals for mates. This may happen at around 5 months of age. They will display dominant signs, such as putting their heads down, pointing their bills up, or spreading their wings out. Sometimes humans that tease or chase geese can provoke this behavior as well. Do not allow children or immature adults to harass your geese.

If a gander does display such behavior toward you, you will need to confront it immediately before it becomes a major problem. First, you should make direct eye contact with the bird. Then, you should step toward the bird. He should back off but if he does not, loosely grab a wing. You will need to work quickly as the gander may try to bite you when you attempt this maneuver. When he tries to back off, let go of the wing. He may try to intimidate you a few times, but it is important for you to confront the gander each time he tries to be the dominant figure. If this behavior continues unchecked, it will turn into a major problem every time you come around the flock.

If you do have an aggressive gander, you can try to get the upper hand by grabbing him and pinning him to the ground. His wings will flap (and they can pack a punch), and he may bite you. He definitely will squawk but keep him down until he submits by resting. Then let him go and repeat as necessary until he leaves you alone. Wearing leather gloves, safety glasses, and a jacket will help protect you from his wings and bill.

Preparing for goslings

Newly hatched geese, called goslings, are purchased directly from the hatchery or through feed stores. You will find more variety in breeds if you order directly from the hatchery. The geese come in a straight-run composed of both sexes, or they can be sexed if you want to have more females or males. Depending upon the hatchery, the price per bird may be more if you order a sexed group, and there may be a minimum number of geese you will need to order.

For the average poultry farmer, a straight-run will be satisfactory for geese raised for meat. Males will be a bit heavier than females at the same age, but female geese make excellent table meat. If you are planning to breed your geese, you will want a ratio of one male to two females. Most ganders will only breed with one or two females.

When purchasing your geese directly from the hatchery, ask when they will be mailed so someone is home to receive them. Your geese will be shipped like chickens and ducks in cardboard containers. They will be shipped as day-old goslings because they have enough nutrients to survive the journey from hatching. Be sure to check that all the geese survived shipping, and if some are dead, submit a claim check to the hatchery.

If you choose to purchase your stock from a store take the same precautions as you would purchasing chicks and ducks from a store. The pen should be clean and dry. They should have food and water, and the goslings should have no eye or nasal discharge. Like chickens and ducks, geese have vents that should be clear of built-up fecal material.

Care of goslings

It is important to set up the pen, feeders, and waterers before your goslings arrive. Like with ducklings, a garage or barn corner that is draft-free and fully enclosed with good ventilation works well. You need 6 inches of space per bird, and this should be increased to 1 foot of space per bird after they are 2 weeks old. For the first two to three days, your goslings will need to be in a smaller confinement area made of cardboard or straw bales. Cover the pen floor with 4 inches of wood shavings, peat moss, or chopped straw for bedding. Maintain this bedding to eliminate wet, dirty spots. Similar to ducklings, the waterer should allow the goslings to dip their heads and bills, and you can add electrolyte or vitamin powder for a boost.

Like with ducklings, a heat lamp with a 250-watt bulb, or a hover-type brooder can be used for heat and light. One heat lamp will supply enough heat and space for up to 20 goslings. Hover brooders typically have instructions for chicks. Since goslings are larger than chicks, brood 1/3 as many goslings as you would chicks. Goslings have the same heat needs as ducklings: A temperature between 85 and 90 degrees Fahrenheit, which can be reduced 5 to 10 degrees every week until the temperature is around 70 degrees Fahrenheit. If you notice your birds huddling under the lamp, you might need to increase the heat. If they lurk at the edges, the heat needs to be lowered. During mild weather, the birds will not need the heat after the sixth week.

Do not let young geese have access to swimming water or leave them outside in the rain because their feathers are not developed. If they have been hatched out by a mother goose, they can have access to swimming water with the adults because the mother will not let them remain in the water for too long, and she will protect them from rain. Goslings can be placed on pasture around 6 weeks of age if the weather is good.

Nutritional requirements

Geese need certain nutrients in their diet to be healthy. These are water, protein, carbohydrates, fats, minerals, and vitamins. *Refer to Chapter 3 for more details on these nutrients.*

Goose rations

Your geese will have specific feeding needs, depending on their age. Different rations include: starter, grower, breeder, finishing, and maintenance.

Starter ration

A good starter feed will set your young geese off to a good start in life. They are generally disease resistant, so a medicated feed is not necessary. In fact, certain medications found in chick starters can cause health problems in geese, such as lameness and even death.

During the first four weeks of life, the goslings should be fed a feed ration of 20 to 22 percent protein. Keep feed in front of geese less than 2 weeks of age at all times. Avoid feeding them a finely ground mash as they may choke on it, and they will waste a lot of the feed. A crumble or pelleted ration is best for the goslings. Starting the first week of life, you can also feed small amounts of fresh growing grass or fresh clippings to the birds. Grit should also be kept in front of the goslings at all times, as this will assist the gizzard in grinding the feed.

Growing ration

After 4 weeks of age, the goose's diet can be supplemented with cracked corn, and it can be switched to a grower ration. An ideal growing ration will have 15 to 18 percent protein. Starting at 6 weeks of age, the geese can be allowed access to pasture and can derive a good part of their nutrition from the grasses. Be sure to keep the grass mowed because long grass can cause death. Cracked corn and grit should continue to be provided to the geese until they are 10 weeks of age. Large breeds of geese that are rapidly growing can eat up to ¾ of a pound of feed a day.

Finishing ration

If your geese are to be slaughtered for home use or for market, they should be fed a finishing ration formulated for turkeys starting one month prior to slaughter. This will provide them extra nutrients allowing them to fatten prior to slaughter. Keep the food in front of the geese at all times. Most geese are slaughtered at 5 to 6 months of age. *See Chapter 5 for information on butchering.*

Maintenance and breeding rations

Birds not intended for slaughter do not need to be fed a finishing ration and instead continue to eat pasture grasses and cracked corn. If pasture becomes poor or the season ends, provide your flock with a maintenance ration of 13 to 14 percent protein along with cracked corn and grit. Full-grown geese should be allowed ½ a pound of feed a day if pasture is unavailable.

If your geese are intended for breeding, they will need to be fed a breeder ration starting five to six weeks before egg production begins. Supply the geese with oyster shells to ensure they get enough calcium to form good eggshells. Keep feed available for the geese at all times.

Husbandry

A plot of pasture enclosed by a 3-foot, woven-wire fence makes a great feed source for the birds when they are about 6 weeks old. As an alternative to heavy, permanent wire fencing, you can use a lightweight, portable electric fence, called poultry netting, made of plastic and electro-plastic strings. By using a fencer to deliver an electric charge to the fence, you can keep the geese inside and thwart predators with a stiff shock if they try to gain access to the flock. Premier fencing (**www.premier1supplies.com**) carries poultry netting, fencers, and plastic PVC posts along with valuable tips on how to use the netting for poultry flocks.

After 6 weeks of age, most of a goose's diet can consist of forage, provided the pasture is in a succulent stage. Geese are great at foraging both bugs and plants. An acre of pasture can support 20 to 40 geese depending upon age and size. The pasture will need to be monitored, as it can quickly become defoliated and heavily soiled by the geese. Grass for goose pasture should be about 4 inches in height. Longer grass can become bound up in the goose's crop – the outcropping of the esophagus – and cause death.

Do not allow the geese to graze the entire pasture down to the dirt. It is best to divide your pasture into five or six separate paddocks and switch the geese to a different paddock frequently. The frequency of moving the geese to a new area depends on the amount of geese in your flock.

Geese do not care for alfalfa or tough, narrow-leaved grasses. Good plants for pasture are brome grass, Timothy, orchard grass, bluegrass, and clover. If you keep your flock on a pasture, provide some feed in covered feeders for the flock. This will supplement their diets when the pastures become grazed down or aged.

You will need to provide some shade for your geese, especially during hot, sunny weather. This can be done by using shade panels or a simple roof made from two-inch by 4-inch, water-resistant plywood. Farm Tek (**www.farmtek.com**) carries many agricultural fabrics and covers that are built to withstand wind and inclement weather. These products make inexpensive shade-protectors and suitable housing materials for older geese.

Water should always be available for the geese to drink and bathe in. If you have a few birds (less than five), you can use a hose and a kid's pool to provide drinking and bathing water. The pool will need to be cleaned and refilled two times a day. If you have more birds, provide multiple waterers and bathing tubs. A small pond can be used for bathing. Geese can quickly damage the pond from their feeding habits and their weight. The banks are particularly vulnerable to damage, so using large rocks to protect the banks will prevent erosion.

Young geese are susceptible to predators. As the larger breeds grow, predators become less of a threat due to the weight of the birds. Dogs and coyotes may kill smaller geese, or they may scare larger geese. When scared, geese will huddle together, and this may cause some geese to be smothered to death. A fenced area with 5-foot-tall fencing makes a good night resting spot for the geese. Be sure to provide your geese with feed and access to a water source in their resting spot. Alternatively, the geese can be locked inside a sturdy building at night to protect them from predators. This is particularly important in geese less than 5 months of age when they are still smaller-sized. Geese younger than 8 weeks of age should not be left out at night on the pasture and should be herded inside a secure shelter for safekeeping from poor weather and predators.

A flock of Canada geese in a pond with a marsh.

Breeding Geese

Like ducks, geese intended for breeding should have a wing or leg band applied so you can identify them when you breed them. Apply the wing or leg bands soon after hatching, and be sure to give each animal only one number. If you order geese directly from a hatchery, they will likely offer to perform this service for you. Records should be kept to note the parents of each hatch, how many eggs each goose lays, if she is broody and for how long, and how many eggs are hatched for each female. Poorly performing geese can be identified by their bands and culled from the flock.

The gender of geese cannot usually be identified by sight alone. They should be purchased as sexed and banded as soon as possible; then it will be easy to identify a gander from goose. For those breeds that cannot be sexed on physical characteristics, vent sexing will need to be done. This is performed when goslings are a couple of days old. It can be done at a later date, but the larger the bird, the more difficult the task. *Refer to Chapter 8 for information on how to sex a goose.*

Only those geese in good physical shape should be kept for breeding. Legs should be straight and free of deformities, as should the beak and wings. A breeding goose should comply with the specific standards of their breed for coloring, body shape, and weight. One gander should only be used for one or two females. Geese should be at least 1 year old when they breed, though some of the larger breeds will not mate until they are 2 to 3 years of age. Geese prefer to mate on water, so if your climate allows for it, keep an area with water for your geese that is deep enough for the geese to swim in. However, since geese start laying eggs in February or March, breeders in colder climates may be unable to provide water for mating.

Allow your breeding flock access to an outside pen, but do not allow them to roam, as you may not find the eggs once they start laying because the females may decide to nest away from the pen. Place nest boxes inside your goose building, allowing one box per female. They should be spaced apart or the females may squabble. Some females may prefer to nest on a self-made depres-

These Canada geese are nesting in a pond.

sion in the floor; if a female does this, make sure you line the pen with clean straw or pine shavings to keep the eggs clean.

Nest boxes for geese should be a minimum of 2 square feet. Larger breeds may require more space. You can build your own boxes out of wood or purchase them from a poultry supply company, although it might be difficult to find a wooden box large enough for geese. If the female incubates her own eggs, make sure she has water and feed available near the nest. A female should not leave her eggs unattended more than once a day. Most females can successfully hatch up to a dozen eggs.

If you plan to incubate the eggs, collect them at least twice daily, but be sure to use caution. During breeding season, the geese can become ornery. You will most likely be hissed at as you collect eggs. To prevent being injured by protective mothers, situate nest boxes near an aisle in the pen, wear gloves, and protect your eyes with safety glasses. You can also create a distraction by feeding the birds as you collect their eggs.

Geese eggs take 28 to 35 days to hatch. Temperature in a forced-air incubator should be 100 degrees Fahrenheit, while in a still-air incubator, the temperature should be maintained at 103 degrees Fahrenheit. The humidity should be at 50 to 55 percent for the first 27 days of incubation. Eggs should be turned over four to six times each day. The final three days before hatching, the humidity level should be increased to 75 percent. When the goslings hatch, the doors to the incubator should be opened to allow the humidity to escape; this allows the goslings' down to dry. After a couple of hours, they should be dry and can be moved to the brooding pen.

Using Geese as Guards and for Weeding Crops

Because of the goose's large size, excellent vision, and loud voice, it makes a good guard bird for the farm. Certain breeds, such as the Egyptian and African, are more aggressive than other gentler breeds, like the Toulouse. Geese are fairly intelligent and can recall people or animals that scare, harass, or frighten them. They also remember troublesome and scary situations. A goose will serve as a guard against intruders. An intruder will surely be deterred from going on your property when faced with a flock of angry, honking, and large geese ready for attack.

Geese also make ideal weeders for gardens and vineyards, as they have a preferential appetite for grasses and will avoid eating broad-leaved plants. Before chemical weed control became commonplace, specialty crop growers relied on geese to keep the grass picked in such crops as asparagus, mint, beets, beans, onions, and potatoes. To use geese for weeding gardens, be sure the garden rows are at least 1 foot apart, build a fence made of poultry netting around the garden, and do not let the geese have access to the garden until your plants are established. With the proper spacing and a fence, a goose will not harm your plants and will not stray from their task.

For plants that ripen above ground, like tomatoes or strawberries, do not allow the geese to weed these plants when the vegetables are ripe. They might peck at the colorful plants or stomp on them. Geese prefer to eat grasses and weeds instead of vegetables or fruits, but they will eat them if the grass becomes scarce. Remove the geese from your garden plot once the weeds have been eaten to keep them from sampling the vegetables or fruits.

Part 4

GAME BIRDS

Game birds provide several purposes, including food, sport, and ornament. Game birds are notoriously difficult to raise due to their temperaments and the fact that they have not been domesticated. Like any wild animal, they are fearful of people and you must be quiet and deliberate when working around these birds. If you are new to raising birds, game birds may not be the best choice. Having a flock of game birds is best left to an experienced bird grower who has some hands-on experience with raising them. However, by raising a small flock, you may find you have just the right temperament to raise game birds and will find yourself with an interesting and rewarding farm project.

Chapter 10

PHEASANTS, PARTRIDGE, AND QUAIL

Pheasants, partridge, and quail are all varieties of game bird. Game birds have played an important part of the human diet and recreation throughout human history. In the past, people who could not afford expensive meat would hunt game birds to supplement their diets of vegetables and starches. At other times, the ruling class has restricted game-bird hunting to their own class and closed off vast hunting preserves to the poor and destitute.

History

Pheasants are native to Asia, but the Romans were responsible for spreading the pheasant to Europe. The bird was first introduced to North America in 1733, but their numbers remained small until 1881. Judge O.N. Denny released around 100 breeding pairs of Chinese ring-necked pheasants in Oregon. From this small release, the pheasant became a common breed and one of the most recognizable game birds in the United States. The ring-necked pheasant is the state bird of South Dakota and an important part of a sizable hunting economy in Great Plain states.

While there are many species of partridge in the wild, only two are commonly raised in the United States by game bird farmers: the Chukar and the Hungarian (also known as the gray). The partridge is native to Europe, Asia, Africa, and the Middle East. It was introduced to the United States in 1889 when Hungarian partridges were brought to Virginia. The Bible mentions the partridge numerous times in the Old Testament. During this time, the bird and its eggs were used as a food source. The partridge has its own place in the famous song "The Twelve Days of Christmas," as it is mentioned as one of the 12 gifts to give on the 12 days of Christmas.

These quail are foraging in the grass.

Quail are classified as Old World quail or New World quail. As the names imply, the Old World quail are native to Europe and Asia. The New World quail is found in North, Central, and South America. The Old World quail consist of 13 species found in Europe, Asia, the Middle East, and Africa. Common species of Old World quail currently raised in the United States include the Japanese (Coturnix) quail and the Button (Chinese Painted) quail. The New World quail is classified in the family Odontophoridae, which includes 31 species of quail. Common varieties raised in the United States include the Bobwhite quail and the California quail. Quail do not perch and are considered ground dwellers that nest, rest, and forage in grasses and hedges.

Game Fowl

Game birds can be a valuable niche product, provided you carefully create a marketing plan prior to stepping into the game bird market. Most of these birds will be sold directly to local consumers or to hunting preserves. They are also raised for showing at exhibitions. If your farm is located near a large metropolitan area, upscale restaurants may provide a market for eggs and meat.

Quail, partridge, and pheasants are also raised for release during the hunting season, although you should contact your local department of fish and wildlife prior to releasing birds into the wild. Most states have strict laws on the release of game birds. Any bird released for hunting should be healthy and have had exercise in a flight pen so they have the best chance of making a great sport bird.

Species of Game Birds

The different species of game birds differ in size, appearance, egg production, and temperament.

Partridge

The partridge is a fairly easy bird to raise with their small size and ground nesting.

Chukar partridge

Male: 21 to 26 ounces
Female: 16 to 19 ounces
Eggs: 40 to 50

The chukar partridge is native to a large region running from India to Greece. They are the national bird of Pakistan. When in the wild, they prefer to live in rocky areas or hillsides. This beautiful bird has orange-red legs, feet, and bill, and grayish feathers. It also has a black band across the forehead, around the eyes, and down the neck. They are fairly calm birds and are easy to raise for hunting release or for meat.

Hungarian partridge

Male: 14 ounces
Female: 13 ounces
Eggs: Up to 100

The Hungarian partridge (also referred to as the gray partridge) is native to Hungary and central Europe. They first became important to hunting in the United States in the early 1900s and are present in sizable numbers in the wild. Male and female Hungarian partridges are similar in appearance. They have brown feathers across the back and gray feathers on the sides and chest. The underside is white and typically has some reddish-brown feathers, particularly on the male. They can fly up to 40 miles per hour.

Quail

Quail are small and plump birds that enthusiasts raise for eggs, meat, and sport.

These quail eggs will provide a delicious breakfast.

Bobwhite quail

Male: 7 ounces
Female: 6 ounces
Eggs: Up to 100 per year

Bobwhite quail are native to the eastern part of North America. The name, bobwhite, comes from the sound it makes when calling. The bobwhite has reddish-brown feathers speckled with black and white spots. The tail is gray. Males have a white throat and a white stripe running from right behind the bill over the eye to the base of the neck. Beneath this white stripe runs a black collar. Females have tan colored throats and stripes. They do not have the black collar as found in the male.

The bobwhite reaches sexual maturity early, at 6 weeks of age, and lays heavy eggs – considering its size. They are only 8 to 11 inches long and are lightweights. They are well-adapted for rearing for meat and egg production.

California quail

Male: 6 to 8 ounces
Female: 5 to 7 ounces
Eggs: 200 to 300 per year

A perched California quail.

The California quail is, not surprisingly, the state bird of California and is a native to the West Coast. It is a unique-looking bird. The breed is defined by a plume of feathers that stick out of the head. In males, this plume is black. Females have a brown plume. The male California quail has more colorful feathers than the females. Males have a dark brown cap surrounding the plume along with a black face with white stripes. The feathers on the back are brown, those on the chest are gray-blue, and the belly feathers are light brown. Females are gray-brown with lighter colored feathers on the belly. They are from 9 to 11 inches in length.

Japanese quail

Male: 4 to 5 ounces
Female: 4.5 to 6 ounces
Eggs: 200 to 300

The Japanese quail has been domesticated in Japan for thousands of years. In the wild, they are found in China, Siberia, Japan, and Korea. They are used for both meat and egg production. This is a prolific egg-laying breed. They usually lay eggs with brown spots. Most quail eggs marketed are from the Japanese quail.

The male Japanese quail is smaller than the female of the species. It has reddish-brown feathers on the throat and breast. In contrast to the male, she has whiter, longer, and more pointed feathers on the throat and chest. Both males and females have brownish feathers splattered with white and black over the rest of the body. Some other color varieties are white, blonde, and tuxedo.

Button quail

Male: Up to 2 ounces
Female: Up to 2 ounces
Eggs: 6 to 12

The button quail, also known as the Chinese painted quail, is the smallest of the quail. In the wild, they are found in India, south China, Indonesia, and Australia. They are about 4 inches long when mature. The male is a dark blue-brown over much of his body with a blue-gray breast and red-brown belly. The face and throat are black and white. Females are brown with splashes of dark colors. Other colors include white, silver, golden pearl, red-breasted, splashed, and cinnamon. This quail is also marketed as a pet, with a lifespan of four years if given the proper care and kept from predators. They make good pets because they are active and do not require much living space. Males are territorial and may need to be separated if they fight too often.

Pheasants

Pheasants are birds that belong to the family Phasianidae. They are large, long-tailed birds that come in a variety of colors and patterns, with the male being more colorful than the female. All pheasants originated in Asia and have been introduced successfully into the wild in North America. There are about 49 species of pheasants.

Pheasants have strong legs and run frequently. Their legs are long and strong, adapted for scratching for food and running. The males engage in battles over territory and typically have spurs on their legs.

Chinese ring-necked

Male: About 2.5 pounds
Female: About 2 pounds
Eggs: 40 to 90 per year

The Chinese ring-necked pheasant is a beautiful bird imported into North America from Asia as a game bird. It is also the most common pheasant, with many subspecies. The ring-necked pheasant quickly established a large population in the wild. Males are showy with metallic red, brown, and green body feathers and metallic blue, green, and black head feathers. A white ring of feathers circles the neck. Females are a dull brown color.

Golden

Male: 3.5 to 4 pounds
Female: 2 to 2.5 pounds
Eggs: 40 to 60

The golden pheasant is an easy pheasant to keep in captivity. It is also a native of China and has been kept in captivity since the early 1700s. It is the most popular of pheasant species kept in captivity. The golden pheasant sports a ruff of feathers around its face and neck, which roosters proudly display during mating rituals. Males are multicolored with scarlet feathers running from the breast to the flanks. The head and neck have tan feathers, and the ruff is golden. The back is green on the upper part, changing to golden toward the rear. Tail feathers are black and red. The female is typically brown with dark highlights. There are mutations, color-wise, of the golden pheasant available. These include yellow-golden and cinnamon-golden. Females have much darker coloring than the males. The species is hardy and it is an easy bird for beginners to raise.

Lady Amherst

Male: 3.5 to 4 pounds
Female: 2 to 2.5 pounds
Eggs: 50 to 60

The Lady Amherst pheasant is in the same genus as the golden pheasant. They are both known as ruffed pheasants, which refers to their head ruffs. The male Lady Amherst pheasant has a black and white neck-ruff. Its back is a greenish color, turning to yellow and red as it approaches the rear. The tail is long with black and white feathers. The female is brown with black stippling on the feathers. The Lady Amherst pheasant is fairly calm and, like the golden pheasant, makes an excellent bird for beginning pheasant raisers.

Green

Male: About 3 pounds
Female: About 2.5 pounds
Eggs: 6 to 12

The green pheasant (*Phasianus versicolor*) is a native of Japan. It has been introduced into the wild in Hawaii and in small numbers on the eastern coast of the United States. The rooster's crown, back, under-parts, and rump are green with some tinges of blue and olive. The male is not aggressive, unlike with the ring-necked pheasants. The green pheasants are amenable to being raised in captivity. The hen prefers to nest under ground cover.

Behavior

For the most part, game birds are shy and easily startled birds. This is because of their status as prey animals and the need to be alert and quickly respond to perceived threats. Partridges and pheasants can fly in short, quick spurts and will likely do so in the pen if they are scared. The pen absolutely needs to be covered with a tight cover to prevent flight losses. It also needs to be constructed so there are no nails, wires, or other protruding items.

Although quail do not fly, they can startle easily and become hurt in pens without a smooth interior. Button quail are also known as popcorn quail because when they are startled they will jump straight up, suddenly, like a kernel of popcorn popping. The tops of quail pens should also be covered to prevent escapes from jumping and to keep predators out.

Nutritional Needs

Game birds can be raised successfully using a game bird diet or a turkey-feeding regime. Just as with other poultry, there are some essential nutrients every game bird needs for life, growth, reproduction, and production. A lack of any of these nutrients will negatively affect the bird's growth, reproduction, and health. These are water, protein, carbohydrates, fats, minerals, and vitamins. *Refer to Chapter 3 for more details on these nutrients.*

Feed companies can give you more information on their particular diet program for game birds. Remember that a turkey diet is perfectly fine for game birds. Some feed companies that provide turkey rations include:

Purina Mills: **www.purina-mills.com**
Nutrena Animal Feeds: **www.nutrenaworld.com/nutrena**

Starter ration

Chicks should be placed on a starter diet immediately after removal from the brooder or purchasing. For the first six weeks, they should be fed a diet that provides them with 24 to 28 percent protein. They should also be provided with plenty of fresh, clean water and grit.

Growing ration

Starting at seven weeks and continuing until 14 weeks, they can be fed a diet of 20 to 22 percent protein. At 15 weeks, the protein diet can be reduced to 20 percent protein until they are marketed.

Breeding ration

The breeding ration for game birds should provide 18 percent protein. The birds should also be provided with crushed oyster shells and grit. They should be placed on the breeding ration along with supplements six weeks prior to breeding.

To make sure your bird's feed is fresh, use it up within six weeks of purchase. Some vitamins can deteriorate after this time and may make your birds susceptible to nutritional deficiencies.

Starting a Flock

If you have purchased game bird chicks, inspect them to ensure all the chicks are alive and healthy. Your brooding pen should be set up prior to placing the chicks into the pen. You will need a heat and light source; a heat lamp will provide your chicks with both needed factors. If your brooding pen is a rectangle or square, the corners should be rounded with cardboard or wire to prevent the chicks from crowding and smothering each other. This is important to remember when working with game birds, as they frighten easily.

All equipment used in the brooder should be disinfected, either with bleach water — use 1 tablespoon of bleach per 1 gallon of water — or with a commercial disinfecting solution. Allow the equipment to dry thoroughly before use. Place 3 to 4 inches of clean bedding on the pen floor. Wood shavings, rice or peanut hulls, ground corncobs, or chopped straw makes acceptable game bird chick bedding. The tops of an open pen should be covered with chicken wire or mesh to keep predators out and, when the chicks feather out, to keep them in.

A 1-gallon waterer should be placed in the pen for every 50 birds. Dip each bird's beak into the waterers when you place them into the brooding pen so they know where to find the water. Watch quail carefully around the waterers, as they are tiny and can fall into the waterer and drown. Placing some clean, small rocks in the waterers' bases the first few days will provide them with footing if they fall into the water.

You can purchase game bird starter feed from local farm stores or from grain elevators, and this should be fed to chicks until they are 6 weeks old. They can also be fed a turkey starter ration. Regardless of the type of starter you choose, it should include 24 to 28 percent protein. After the chicks are 4 weeks old, you can add some whole grains to the diet, as long as you also provide the birds with grit.

From 6 weeks until they are 14 weeks, a game bird grower or turkey grower ration with 20 percent protein should be fed to the birds. After this, they can be fed a ration with 15 percent protein. Twenty inches of feeding space should be allowed per 100 chicks. Scatter some feed on pieces of cardboard or newspaper near the feeders the first few days so the chicks can find the feed.

The heat source should be lowered so it will be 2 inches over the chicks. Ideally, the temperature under the heat source should be 95 degrees Fahrenheit. This temperature should be maintained for the first week; then the heat can be lowered by 5 degrees each week until it reaches 70 degrees Fahrenheit. During the first few days, you may want to confine the chicks close to the heat source, feed, and water with a small cardboard or wire ring to prevent them from wandering into the cooler parts of the brooding pen. A light should be on continuously during the first week; starting the second week, the light can be left on for 12 hours a day until the chicks are ready to be removed from the brooder. If the pen starts to smell of urine – an ammonia-like smell – or if there are any wet or excessively dirty spots, the bedding should be removed and replaced with clean litter.

Housing the Game Bird

When the birds are fully feathered – usually around 6 weeks of age – they can be moved to growing cages or a flight pen. Partridges and quail can be kept in inside pens if being raised for food, but if they are raised for hunting, they should be placed in a flight pen. Inside cages for these two species, you should allow for 1 square foot of floor space for quail and 1½ square feet floor space for partridges. Other species should only be raised in flight pens. Given the button quail's small size, you will need to make sure the bird cannot escape from the pen's wires.

Construct the flight pen in a long, rectangular shape. It should be located in a quiet area of the farm, away from human and vehicle traffic. Plan accordingly to allow 2 to 4 square feet per partridge or quail, or 10 square foot per pheasant. The pen should be built against a building to allow the birds to have access to the inside in case of poor weather. Make the sides with upright wooden or steel posts and chicken wire. The wire should be sunk 18 inches into the ground along the base of the pen to prevent predators from tunneling into the pen. Cover the top with poultry netting to keep the birds from flying out, and build a wall 24 inches tall made of boards around the pen's perimeter to protect your birds from the wind.

Vegetation should be planted inside the pen to provide cover for the birds. Vegetation provides environmental enrichment and protection for birds that are being picked on by other birds. This can range from a few conifer trees to annual grasses like millet, oats, and wheat.

Waterers and feeders should be scattered throughout the pen. Feeders should be covered to prevent moisture from damaging the feed.

Breeding

Game bird eggs can be purchased from breeders and incubated much like chicken eggs.

Egg incubation

Eggs to be incubated should be clean, free from cracks, and not abnormally-shaped. If you collect the eggs yourself, collect them three times a day to prevent them from breaking and soiling. Any heavily soiled eggs should not be used for incubation.

The incubator should be cleaned and disinfected before you use it and after the eggs have hatched. After it is cleaned, it should be set at the desired temperature and humidity and running 24 hours prior to placing the eggs inside. Always read the instructions that come with your particular brand of incubator, especially on how to ventilate the incubator and read temperature and humidity levels.

The eggs should be turned at least three times daily. Marking the eggs with a pencil or crayon will help you remember which sides of the eggs were last turned. As an alternative to hand-turning, especially if you have a large number of eggs to incubate, you may want to invest in an incubator with an automatic egg turner. When the chicks start to hatch, the incubator temperature should be decreased to 95 degrees. Remove chicks from the incubator when the hatched chicks are dry and running around the incubator.

Incubator Type	Incubation Temp.	Humidity	Hatch Temp.
Still Air	102 to 103°F	60 percent	100 to 101°F
Fan	99.5 to 100°F	60 percent	98.5°F

Incubation Times

Species	Days
Chukar Partridge	23 to 24
Hungarian Partridge	24 to 25
Bobwhite Quail; California Quail	23 to 24
Japanese Quail	17 to 18
Button Quail	16 to 17
Pheasant	22 to 25

Breeding game birds

A healthy game bird chick starts with healthy parents. Breeding stock should be free from disease and any abnormalities. They should have the body shape, size, and color pattern that meets their breed standards. Breeders can be held in wire group-pens or in flight pens. Provide covered nest boxes for the birds; pheasants should be allowed a nest box 2 feet wide by 6 feet long by 1 foot high. This will provide enough nesting space for up to 24 females. For the smaller birds, the dimensions can be halved. Provide a layer of nesting material — straw or shavings — and check it daily for soiling or dampness when collecting the eggs.

Remember to feed your breeders a laying ration along with plenty of crushed oyster shells and grit.

Despite your best attempts, you may find some eggs do not hatch. The reasons for a low or reduced hatchability vary, but some common reasons are: infertile eggs, incubating old eggs, fluctuating temperatures prior to incubation, dirty eggs, not turning the eggs enough, cracked eggs, and incubator problems, including poorly regulated temperature and humidity.

The California blue quail would make an exotic addition to your farm.

Chapter 11
ORNAMENTAL GAME BIRDS

Like their name suggests, ornamental game birds can add a charm to your farm with their colorful feathers and personalities. However, as you will learn, ornamental game birds provide practical purposes as well.

History of the Guinea Fowl and Peafowl

Guinea fowl and peafowl are usually kept in the United States as ornamental birds. Both species have unique qualities that make them a welcome addition to the farm. Guineas and peafowl are raucous and protective of their territories, making them ideal barnyard alarm systems. They belong to the same family as pheasants and partridges.

The guinea fowl is descended from wild guinea fowl from Africa, specifically the country of Guinea. In the wild there are six varieties, but only three have been domesticated in the United States. Guineas have been domesticated for centuries and were used as a prized food source by the ancient Greeks and Romans. The meat of the guinea is perfectly fine to eat, although that is an uncommon practice in the United States.

Many people keep guineas to keep the insect and tick populations around the farmyard under control. In areas where Lyme disease is prevalent, guineas have been used to control the ticks that carry this debilitating illness. Guineas are considered omnivores because they eat vegetation but will also seek and destroy small reptiles and snakes.

They will roam through gardens, leaving flowers and vegetables alone, in their quest to eat insects and weed seeds. In fact, their eating patterns are different from chickens; they do not scratch the dirt as chickens do, but instead leave the soil and plants unscathed. During the night, they should be kept in pens, as owls, hawks, and other predators will attack guineas.

Peafowl are ornate birds with long, showy tail feathers. The most familiar peafowl is the peacock. The males have tail feathers characteristic of this breed. They originated in Asia, where they are still found in the wild. Interestingly enough, Alexander the Great brought the peacock to the areas he controlled.

The peacock even holds a place in Greek mythology. According to the myth, Zeus, the Greek god of the sky, transformed one of his lovers, Io, into a heifer to protect her from his jealous wife Hera, the goddess of marriage and love. Hera, in turn, tricked Zeus into giving her the heifer. Hera placed Argus, a giant with 100 eyes, in charge of watching the heifer. In turn, Zeus sent Hermes, the god of herds, to free the heifer. Hermes made Argus fall asleep by talking for hours and hours, which caused him to close all his eyes. Hermes then killed Argus to free Io. When Hera discovered Argus was dead, she took Argus's eyes and placed them on the peacock's tail to honor her faithful servant.

In India, the peacock is the national bird of the country. The Hindus believe the peacock is sacred, and hunting them is illegal in India. When the god of thunder and war is portrayed, a peacock is used. Hindu legend also believes that peafowl are able to charm snakes.

In addition to their colorful history, peafowl make a colorful addition to the farmstead. They are happiest if they are allowed to roam. Their routine is to forage around dawn and dusk. Trees make ideal roosting places for peafowl; some have even been known to roost on the tops of farm buildings and houses. They are incredibly noisy at night, so peafowl roosting areas should be located away from the house. One last point about peafowl is that they are able to live for over 50 years, which is an important consideration when deciding to add peafowl to your farm.

Guinea Fowl

Male: 4 pounds
Female: 3½ pounds

As with many other species of poultry, breeders and fanciers have toyed with genetics to produce the many different colored guineas now available. Some of the more popular, yet rare, colors include buff and porcelain. One note: Plan well in advance if you have your heart set on a particular color pattern. Contact a breeder at least six months prior to hatching (usually in the spring) to inquire about ordering keets — baby guinea fowls. The rarer or more popular colors may have a limited availability.

Helmeted guinea

These helmeted guinea fowl are foraging.

The helmeted guinea's defining feature is a helmet made from a bony ridge on the head. This is the most common type of guinea raised on farms and the one that comes to mind when people mention guineas. The helmeted guineas come in many colors: light tan to brown; blue to gray; lilac to purple; and pure white. The most common color types of the helmeted guinea fowl found are as follows:

- **Pearl:** This variety has purple-gray feathers with small white dots, or "pearls."
- **White:** As its name suggests, the white guinea has pure white plumage. As opposed to the colored breeds, the skin and meat of the white guinea are light colored. This makes them more visually appealing as a table meat bird.
- **Lavender:** The lavender guinea has lighter gray/purple feathers than the pearl variety. It also is splashed with small white dots.

Crested guinea

The crested guinea has black feathers arranged in a topknot. Crested guineas are hard to find in the wild because they have experienced significant habitat loss. These birds are dark black to gray and have white spots covering these feathers. The breed is named for the black feathers sticking out of the top of their heads. Some growers have reported that the crested guinea can be aggressive towards humans. Due to this fact, the breed is best raised by experienced poultry farmers.

Vulturine guinea

An up-close view of a guinea fowl's head.

The vulturine guinea originated in Africa. The vulturine guinea has a blue skin on its head and neck, red eyes, and bright blue feathers smattered with spots and stripes. They are curious, yet excitable and need plenty of pen space – about 100 square feet per bird – to be successfully raised. This breed is tame in captivity, but in the wild they travel in large flocks. They are intolerant of cold, and if temperatures dip below 40 degrees Fahrenheit, they will need an insulated building with supplemental heat.

A great website for more information about guineas is the Guinea Fowl Breeders Association. This fine group has a vast online wealth of information on guineas: **www.gfba.org**.

Peafowl

Male weight: 8 to 14 pounds
Female weight: 6 to 9 pounds

The most common varieties of domesticated peafowl are the Java and the Indian. A male is called a peacock, and a female is called a peahen. Together, the two are called peafowl. Although these birds are beautiful, it is not advisable to raise them with other breeds.

Indian

The Indian peacock, also known as the blue peacock, is probably the most recognizable peafowl variety. This breed can live up to 20 years in the wild. It has a long tail-feather that is about 5 feet long. The train makes up about 60 percent of the Indian peacock's body length, and it consists of 125 to 150 feathers, each with an eye-spot, or ocellus. The peacock fans his tail when trying to impress the females. The female Indian peacock has less-pronounced coloring. She has a crest and brown feathers with a white belly and green neck. The Indian peacock is sensitive to cold weather and should be put in an insulated and heated pen in cold weather.

Java

The Java peafowl is also known as the green peafowl. The breed is native to the island of Java. They are not as common as the Indian peafowl in the wild or as domestic fowl. They sport a crest of distinctive feathers at the back of the head. The peacock has blue skin around the face, a metallic green head and neck, blue-green shoulders, and a dark green abdomen. It is larger and leaner than the Indian peacock, and its overall length from beak to the end of the tail can reach over 7 feet long. Peahens are also colorful but more muted than the peacock; usually they have a dull green color. Growers report that the Java peafowl is flightier and more aggressive towards humans than the Indian peafowl. The Java peacock also needs more careful care, as it is intolerant of cold weather. It will require an insulated and heated pen when temperatures go below 40 degrees Fahrenheit.

Other peafowl varieties

While the Indian and Java peafowl are the most well-known peafowl, breeders and enthusiasts have developed many color varieties of peafowl. One of the best-known color varieties is the white peafowl. This is not an albino bird, but rather a color mutation of the feathers. Both the male and female of this variety are white with blue eyes, with the chicks hatching out as yellow.

Green is another variety of color. The Burmese green is found in the wild in Burma, and it is a duller green than the Java. The Indo-Chinese green is another green-hued peafowl found in the wild. It is a little less brilliantly colored than the Java peafowl.

Pied peafowl are peafowl speckled and dashed with white feathers. Some varieties of pied peafowl are the India blue pied, the silver pied, Spalding pied, and the India blue pied white-eyed, with "white-eyed" referring to the eyes on the tail feathers. Pied-colored peafowl do not necessarily breed true; there is no guarantee that a mating of a pied peacock with a pied peahen will produce pied chicks.

Black-shoulder peafowl have feathers in the shoulder area that are black with a blue-green hue. There are many color variations involving the black-shoulder color pattern: black-shoulder pied, Spalding pied, opal black-shoulder, and bronze black-shoulder Spalding.

Peafowl also come in a few other general colors that also combine with the pied, black shoulder, and white-eyed patterning. These include peach, opal, purple, charcoal, midnight, bronze, and cameo. A great in-depth website with full colored photos of these and other color varieties can be found at Amy's Peacock Paradise (**www.amyspeacockparadise.com**).

There is one other species of peafowl that should be mentioned, although it has never really been domesticated. The Congo peacock, a native of Africa, is kept in zoo wildlife displays. This is a smaller and less vividly colored peacock that is absolutely cold intolerant and needs specialized housing in temperate climates.

An excellent website for information about peafowl is the United Peafowl Association (**www.peafowl.org**). This site gives excellent information on peafowl from general care to a breeder directory.

Nutritional Needs

Guineas and peafowl can be raised successfully using a turkey-feeding regime. Guineas can also tolerate the same diet as chickens. Some guinea owners will raise their guineas with chickens, allowing them to eat the same foods. Just as with other poultry, there are some essential nutrients every guinea and peafowl needs for life, growth, reproduction, and production. A lack of any of these nutrients will negatively affect the bird's growth, reproduction, and health. These are water, protein, carbohydrates, fats, minerals, and vitamins. *Refer to Chapter 3 for more details on these nutrients.*

Feed companies can give you more information on their particular diet program for guineas and peafowl. Remember that a turkey diet is perfectly fine for guineas and peafowl. Some feed companies that provide turkey rations include:

- Purina Mills: **www.purina-mills.com**
- Nutrena Animal Feeds: **www.nutrenaworld.com/nutrena**

Keets and chicks should be provided with a well-balanced turkey feed ration as soon as they hatch. If you plan to raise your peafowl on the same farm as chickens, a blackhead preventive should be used in the water, as they are susceptible to this disease spread by chickens. Always provide your guineas and peafowl with grit and oyster shells at all times.

Carefully monitor your keets and chicks, checking them many times a day to make sure all the babies are eating and drinking. To understand the health of your flock, observe your newly hatched keets and chicks for about 20 to 30 minutes at least three times a day during the first three days. Any sick-looking or dead birds should be promptly removed from the flock. Dead birds should be properly disposed of in the garbage or through incineration.

Starter ration

A good starter ration will provide 28 percent protein. This high amount is essential to proper growth and development, which is crucial during the first few weeks when the keets and chicks will be growing rapidly.

Grower ration

Starting at 2 months of age, the keets and chicks can be gradually switched over to a growing ration. The feed should have a 24 percent protein level. If possible, they should also be allowed access to pasture. The birds will be old enough to forage for green growing grass or alfalfa. Cracked corn, millet, or oats can also be offered in a separate feeder. Continue to provide grit and oyster shells as well.

Breeding ration

Adult guineas and peafowl intended for breeding should be fed a turkey ration with 16 percent protein. Supplement this ration with small amounts of alfalfa (hay or fresh) or access to grass, and continue to provide grit and oyster shells.

Behavior

As mentioned, game birds are not domesticated like chickens, turkeys, and waterfowl, so some special considerations will need to be made.

Guinea fowl

Guineas are a bit shyer than chickens and, if not tamed, can be susceptible to panic when approached by people or animals. To counteract this tendency, you can try to tame your guineas beginning when they are day-old keets. Interact with the keets calmly when observing, feeding, and watering. Talk quietly to them and gently pick them up and stroke them. Consistently doing this each time you visit your brooding area will give you a calmer flock that does not fear you when you approach.

Since guineas do fly, you may want to clip their wings to keep them from flying out of their pens and to keep them out of trees. Clipping involves taking a sharp scissor and cutting off half the length of the primary flight feathers; these are the last ten feathers on the wing. You should only cut the primary flight feather on one wing. When the bird molts, these feathers will re-grow, so you will need to repeat the process after each molting session. This may mean re-clipping the wing every few months for younger birds or once a year for older birds. *Refer to Chapter 3 for more information on clipping wings.*

Peafowl

Peafowl are noisy birds, a fact you need to consider when choosing a location for your flock. Buildings and pens should be located as far as possible from neighbors, as the calls and cries of peafowl can be ear-piercing. In fact, the screams of a peacock can sound like the screams of a woman in distress.

Unlike many other domesticated poultry, the peafowl is capable of extended flight, but generally they will stay around the area where they were hatched or brooded. If you have a good stand of trees near your peafowl buildings, you will likely find that these trees will become your peafowl's roosting area.

Egg Laying and Brooding

Many people only think of chickens as a source of eggs, but guinea fowl can provide you with large, tasty eggs as well.

Guinea fowl

Reproduction

Guinea eggs can be eaten just like chicken eggs. They are much larger though, so adjust recipes accordingly. One guinea egg equals approximately two chicken eggs. If you plan to breed your guineas, one male will mate with five to six females. In the wild, male guineas prefer to mate with only one female. Guineas reach maturity around 7 months of age. A hen has the potential to lay between 80 to 150 eggs a year.

Guineas used for breeding should be penned. Hens like to make nests in grassy areas, so you will need to cover the pen or clip their wings. Guinea hens are not attentive to eggs and will lay their nests where predators can find the eggs or newly-hatched keets. They will also lead their keets through wet grass. Because the keets have not developed their feathers, becoming wet and chilled will cause them to die from exposure.

Pens and equipment for breeding guineas should be cleaned and disinfected prior to placing the breeding flock into the pen. Put a generous layer of shavings, chopped straw, or sawdust down. You can use chicken nest boxes for the hens to lay the eggs. If you plan to incubate the eggs, eggs should be collected twice a day and gently cleaned to remove any dirt or manure.

Incubation of eggs

Guinea eggs are incubated much like the eggs of other poultry species. You will only want to incubate eggs that are clean and normal-shaped. Never incubate eggs that are soiled or cracked, as this will allow disease to enter into the egg and kill the developing embryo. Whether you purchase fertile eggs or collect the eggs from your guinea flock, it is best to let the eggs settle for 24 hours prior to placing them in the incubator. This allows the egg contents to arrange themselves in proper position.

Guinea eggs can be stored for up to one week prior to incubation at 65 degrees Fahrenheit. Store the eggs in a wicker basket lined with a cotton or flannel material. This will allow oxygen to permeate the eggs and reach the developing embryo. Turn the stored eggs once a day. Four to six hours before loading them in the incubator, bring the eggs to room temperature to prevent heat shock to the eggs.

Purchase the incubator a few days before you plan to use it and read the instructions that come with your particular brand of incubator. Make sure you understand how to ventilate the incubator and read temperature and humidity levels. Clean and disinfect the incubator prior to use with a mild bleach solution, even for first-time use of a new incubator. Set the incubator at the desired temperature and humidity and let it warm up for 24 hours prior to placing the eggs inside. The eggs will need to be incubated for about 28 days. Guinea eggs should be incubated at 99.5 degrees Fahrenheit and 65 percent humidity for the first 25 days. Then the temperature is decreased to 98.5 degrees Fahrenheit and humidity is increased to 80 percent for the final three days of hatching.

The eggs should be turned at least three times daily. Lightly marking the eggs with a crayon or pencil will help you remember which sides of the eggs were last turned. When the chicks start to hatch, do not be tempted to help them out of their shells. They need to perform this act on their own to get off to the proper start. Remove keets from the incubator when the keets are dry and place them into the prepared brooding pen.

Brooding

If you have purchased keets through a mail-order hatchery, inspect them to ensure all the keets are alive and healthy. Your brooding pen should be set up prior to placing the keets into the pen. You will need a heat and light source; a heat lamp will provide your keets with both needed factors. If your brooding pen is a rectangular or square, the corners should be rounded with cardboard or wire. Keets tend to congregate in corners and might accidentally smother one another. Poultry supply companies or hatcheries will often sell rolls of cardboard specifically for this purpose.

Prior to use, the brooder and any other equipment should be disinfected, either with bleach water (1 tablespoon of bleach per gallon of water) or with a commercial disinfecting solution prior to ordering your keets. Allow the equipment to dry thoroughly before use. Place 3 to 4 inches of clean bedding on the pen floor. Wood shavings, rice or peanut hulls, ground corncobs, or chopped straw make acceptable keet bedding. The tops of an open pen should be covered with chicken wire or mesh to keep predators out and, after the keets feather out, to keep them in.

A 1-gallon waterer should be placed for every 50 birds. Dip each bird's beak into the waterers when you place them into the brooding pen so they know where to find the water.

Twenty inches of feeding space should be allowed per 100 keets. Scatter some feed on pieces of cardboard or newspaper near the feeders the first few days to allow the keets to find the feed.

The heat source should be lowered to hang 2 inches over the keets when they are standing their tallest. Ideally, the temperature under the heat source should be 95 degrees Fahrenheit. This temperature should be maintained for the first week; then the heat can be lowered by 5 degrees each week until it is 70 degrees Fahrenheit. During the first few days, you may want to confine the keets close to the heat source, feed, and water with a small cardboard or wire ring to prevent them from wandering into the cooler parts of the brooding pen. A light should be on continuously during the first week; starting the second week, the light can be left on for 12 hours a day until the keets are ready to be removed from the brooder. If the pen starts to smell of urine (ammonia smell) or if there are any wet or excessively dirty spots, the bedding should be promptly removed and replaced with clean litter.

Peafowl

Breeding peafowl can be a rewarding hobby. These birds will make a great addition to your farm, or you can sell them to make a profit.

Reproduction

Peafowl take longer than other domestic poultry to reach breeding maturity; this usually occurs at around 2 years of age. One-year-old peahens will occasionally lay eggs during the summer, but normally peahens do not reach maturity until the 2-year mark. Peacocks are not monogamous and will mate with more than one female. One male can successfully mate with up to five females. The male will help defend the peahen's nest and help her raise the young chicks as well. The peacock also will form a strong bond with his chicks as they grow older, and peacocks will show chicks how to forage for feed and roost.

Hens prefer to nest in hedges or dense shrubs. However, allowing your peafowl to nest outdoors will expose her and her eggs to predators. Keeping the hens confined to a building plus allowing her access to an outside run will keep her reasonably happy. Provide plenty of nesting material (deep straw or shavings) for her to make a nest. Young hens may refuse to incubate eggs in confinement. If so, the eggs can be incubated much like any other poultry eggs. Hens that will incubate eggs can successfully incubate ten eggs at one time.

After the mother hatches her clutch, it is important to keep the mother and chicks in confinement to prevent death loss of the chicks to poor weather and predators. The hen and chicks should be kept confined for six weeks until the chicks are fully feathered. Provide the chicks and mother a growing ration during this time period along with plenty of fresh water, oyster shells, and grit.

Pen cleanliness is vital during this period. Remove soiled or wet bedding as soon as it is noticed and replace it with fresh material. Keep feeders and waterers clean. If they become soiled with manure, scrub and disinfect them before refilling with water or feed. The pen should also be well-ventilated to minimize the chance of any respiratory illness.

Incubation

If you wish to increase your chick numbers, remove eggs from the nest and artificially incubate them. Remove the eggs daily and lightly clean any soiled eggs with warm water. The hens will continue to lay eggs, usually one every other day. You can expect your hen to lay as many as 30 fertilized eggs.

Peafowl eggs are about the size of turkey eggs. The eggs you incubate can be stored at 65 degrees Fahrenheit for up to seven days prior to incubation. When storing the eggs, turn them once a day. Only eggs that are clean, free from cracks, and not abnormally shaped should be incubated. Always read the instructions that come with your particular brand of incubator, especially on how to ventilate the incubator and read temperature and humidity levels.

The incubator should be cleaned and disinfected; then set at the desired temperature and humidity. Incubate peafowl eggs at 99 to 100 degrees Fahrenheit. Record the temperature of different spots in the incubator to make sure it is consistent. The humidity level should be 60 percent or a wet bulb temperature of 86 degrees Fahrenheit. Humidity is adjusted by filling the water troughs provided with the incubator. Placing the incubator in a temperature-controlled room will keep the temperature inside the incubator constant as well. Do not to place the incubator near a heating vent or in direct sunlight. Allow the incubator to run for 24 hours prior to placing the eggs inside.

The eggs should be turned at least three times daily. Marking the eggs with a crayon or pencil will help you remember which sides of the eggs were last turned. When the chicks start to hatch, the incubator temperature should be decreased. Remove chicks from the incubator when the hatched chicks are dry and place them into a prepared brooding pen.

Brooding

If you have purchased chicks via a mail-order hatchery, inspect them at the post office or in front of the mail carrier to ensure all the chicks are alive and healthy. The brooding pen should be set up with feed, water, and lights in place prior to placing the chicks into the pen. You will need a heat and light source; a heat lamp will provide your chicks with both needed factors.

If your brooding pen is rectangular or square, the corners should be rounded with cardboard or wire. It is common for the chicks to congregate in corners and crowd one another – this can lead to smothering. Poultry supply companies or hatcheries will often sell rolls of cardboard designed specifically for this purpose.

All equipment used in the brooder should be disinfected, either with bleach water (1 tablespoon per gallon) or with a commercial disinfecting solution, prior to ordering or placing your hatched chicks into the pen. Allow the equipment to dry thoroughly before use. Place 3 to 4 inches of clean bedding on the pen floor. Wood shavings, rice or peanut hulls, ground corncobs, or chopped straw are all good bedding material for chicks. The tops of an open pen should be covered with chicken wire or mesh to keep predators out of the pen.

A 1-gallon waterer should be placed for every 25 birds. Dip each bird's beak into the waterer when you place them into the brooding pen so they know where to find the water.

Twenty inches of feeding space should be allowed per 50 chicks. Scatter some feed on pieces of cardboard or newspaper near the feeders the first few days so the chicks can find the feed.

The heat source should be lowered to hang 2 inches above the height of your chicks. Ideally, the temperature under the heat source should be 95 degrees Fahrenheit. This temperature should be maintained for the first week; then the heat can be lowered by 5 degrees each week until it is 70 degrees Fahrenheit. During the first few days of brooding, you may want to confine the chicks close to the heat source, feed, and water with a small cardboard or wire ring to prevent them from wandering into the cooler parts of the brooding pen. A light should be on continuously during the first week; starting the second week, the light can be left on for 12 hours a day until the chicks are ready to be removed from the brooder. If the pen starts to smell of urine (ammonia smell) or if there are any wet or excessively dirty spots, the bedding should be promptly removed and replaced with clean litter.

Housing for mature peafowl

Mature peafowl can be allowed to roam the farmstead during good weather. During harsh weather, or when hens have newly-hatched chicks, they should be provided shelter. Since peafowl are able to fly and need to roost in confinement, the ceiling should be at least 8 feet tall. It should also be a large building, providing each mature bird 10 square feet or more of floor space. The birds should be kept off the floor; it is advisable to use ¼-inch welded wire on the coop's floor to keep the birds' feathers in good, clean condition. Outside pens or runs should be constructed of fences that are 8 feet tall to prevent the peafowl from escaping. Provide roosts for your peafowl similar to roosts for turkeys.

Chapter 12

POULTRY HEALTH

An essential part of having a successful flock is being knowledgeable regarding health issues that can affect your birds. This topic has its own chapter because the different species of poultry discussed in this book — chickens, ducks, geese, game birds, and ornamental game — contract many of the same diseases. This chapter is divided into sections that describe common viral diseases, bacterial diseases, internal parasites, external parasites, nutritional deficiencies, and behavioral problems that can plague your birds. Underneath the name of each ailment is the type of poultry the ailment usually affects.

Disease conditions

Many of the diseases of poultry are diagnosed through necropsy of dead birds or laboratory testing of blood, secretions, or tissue samples. Viral diseases cannot be cured by use of antibiotics, but most bacterial diseases can be treated with antibiotics. Nothing is a better treatment than purchasing disease-free birds and placing them in a clean and sanitized environment. Unfortunately, many diseases strike swiftly, killing the birds before the owner realizes there is a problem. Even if the poultry survive the initial illness, it is common for the survivors to be unthrifty and, in some cases, chronically infected with the disease.

Viral diseases

Avian influenza

Affects: chickens, turkeys, ducks, geese, game birds, ornamental game

This can cause a wide range of signs in poultry ranging from a mild disease with few deaths to a fatal, rapidly spreading disease. Waterfowl are more resistant to avian influenza than other poultry breeds. In fact, waterfowl can serve as a source of avian influenza to turkeys and chickens, causing serious disease in these two species. The signs of avian influenza are variable depending upon the strain of virus, but signs include sneezing, coughing, ruffled feathers, lower energy, diarrhea, swollen head, or sudden death. There is no specific treatment or vaccination. Prevention revolves around keeping wild birds away from domestic poultry flocks. Ducks can develop immunity to the virus, serve as carriers, and spread it to chickens, turkeys, geese, and even humans.

Newcastle disease

Affects: chickens, game birds, ornamental game

This virus causes a fast-spreading, fast-acting disease of the bird's respiratory system. Newcastle disease can cause high death losses depending on the virus type. Infected birds spread the virus in their droppings and through droplets from the respiratory tract. It can also be spread through dust contaminated with the virus.

Depending on the viral strain, the disease can be mild, moderate, or severe. If the virus type does not cause immediate death, sick birds will show such signs as lack of appetite, weight loss, loss of coordination, difficulty breathing, sneezing, diarrhea, and nervous system signs.

The diagnosis can only be made via laboratory testing, and in the United States the disease needs to be reported, due to its rapid spread and potential for high death losses among poultry flocks. There is no specific treatment for Newcastle disease; prevention is best through vaccination and good biosecurity. Hatcheries can vaccinate young chicks against this virus, or vaccines can be given in the water or as a spray.

Fowl pox

Affects: ducks, game birds, game birds, ornamental game

This virus is spread by direct contact with infected birds or through mosquitoes and other biting insects. It affects the conjunctiva around the eyes, the throat, and other unfeathered parts of the bird. This is a slow-spreading disease and only a few birds in the flock will have visible lesions at any given time. The presence of fowl pox is evidenced by a raised scab with a crater-like center. There are two forms of fowl pox: the wet form and the dry form. The wet form causes canker-sore-like lesions in the mouth and throat, which can cause trouble breathing due to obstruction of these respiratory passages. The dry form causes raised, bumpy growths on the legs, which can cause problems with growth and egg production. Treatment consists of treating the pox lesions with an antiseptic. Vaccination and insect control are the main forms of prevention.

Hemorrhagic enteritis

Affects: game birds, ornamental game

This virus causes an extensive inflammation and subsequent hemorrhage of the intestines. Young birds between 4 and 12 weeks of age are the most susceptible to hemorrhagic enteritis. Signs include bloody droppings, depression, and paleness due to blood loss. The disease is spread when the bird consumes feed, water, or litter contaminated with fecal material containing the virus. Like with other viruses, there is no specific treatment for hemorrhagic enteritis, although the use of antibiotics in the water will help to prevent any secondary bacterial infections. Good husbandry and sanitation are keys to preventing hemorrhagic enteritis, along with vaccination.

Chlamydiosis

Affects: ducks

Ducks are susceptible to chlamydiosis, or parrot fever. Signs of an infected duck include nose and eye discharges, sinus infections, reddened eyes, diarrhea, weight loss, and loss of appetite. The disease is spread from infected birds to healthy birds through discharge and feces. Wild birds can spread the disease to domestic ducks. Chlamydiosis is also spread through contaminated boots, clothing, and equipment. Once an infected duck recovers, it can still be a carrier of the organism. The duck can spread the disease to other birds and humans. If a bird is sick, it needs to be isolated from the flock. Treatment is through the use of the antibiotic chlortetracycline. Be sure to wear protective clothing, gloves, and even a facemask when treating a sick bird. Keeping wild birds away, disinfecting the pen, replacing the cage paper, and examining ducks for signs of illness before purchase are all ways to prevent parrot fever.

Infectious hepatitis

Affects: ducks

This disease affects young ducklings 2 to 3 weeks old. A virus that is either ingested or inhaled causes infectious hepatitis. The sick duckling appears to be unable to gain its balance and may lie on its side with its head drawn back toward the tail. Their legs will also make paddling motions. Once a duck is infected, it can die within the hour and will likely die within the day. To stop the spread of the disease, vaccinations are available for healthy ducklings in an infected flock. To prevent illness, mothers can be vaccinated two weeks prior to laying eggs to pass immunity on to the ducklings. Separating ducks by age will also prevent an infectious hepatitis outbreak.

Marble spleen disease

Affects: game birds

Pheasants can be affected by marble spleen disease, a viral disease common in confinement-raised pheasants. The virus is closely related to a turkey virus that causes bloody diarrhea in turkey poults. Despite its name, marble spleen disease is considered to be a respiratory disease. If noticed in time, the symptoms are difficulty breathing and a depressed appearance. Most birds are found dead before signs are noticed. Pheasants less than four weeks of age are resistant to the disease due to the presence of antibodies from their mothers. After this time, these antibodies are no longer in their systems, which then leaves the pheasants susceptible to infection until their own immune systems kick in.

The virus is spread through equipment or footwear contaminated with soil or feces. The pheasant picks up the virus through the mouth, eventually leading to a system-wide infection. It is spread to other birds via the droppings of infected birds. Death rates can reach up to 60 percent, but they are more typically around 15 percent. Even if the pheasants survive the initial viral infection, they can become run-down and susceptible to secondary bacterial infections.

A necropsy shows an enlarged spleen with a mottled color along with fluid build-up in the lungs and other internal organs. Your flock can be given a vaccination to prevent infection. This vaccine is given in the water to pheasants when they are 4 to 5 weeks of age. To prevent this disease, you will need to practice good hygienic measures, especially if the people working with the pheasants have been in contact with pheasants not located on the farm.

Quail bronchitis

Affects: game birds (quail)

Quail are susceptible to a viral respiratory disease called quail bronchitis. Young quail are most severely affected; coughing, sneezing, and wheezing are the predominant signs. There is no vaccination available for this disease, which can cause up to 100 percent of the flock to be lost. To help prevent this disease, keep wild birds and rodents away from the flock through pen and building maintenance. Great nutrition and scrupulous pen and equipment disinfection will also help minimize the chance of an outbreak of quail bronchitis.

Bacterial diseases

Staphylococcosis

Affects: game birds, ornamental game

Staphylococcosis is caused by the Staphylococcus bacteria. This bacterium is found on normal birds' bodies, but wounds, navels, and cuts can become infected with Staphylococcus and the bacteria can spread throughout the body. It can cause infection in almost any body part, including the joints, navel, lungs, and liver. The most common site and signs of infection occur in the joints, with lameness being the prominent sign. Diagnosis depends upon culturing and identifying the type of the organism in the lab. To treat the disease, your bird will need antibiotics, but be sure to get the right type because the bacteria can be resistant to many antibiotics. This bacterium can also be spread to humans, so careful handling and sanitation is important if you suspect your bird has this disease.

Mycoplasmosis

Affects: chickens, game birds, ornamental game

This bacterial disease is caused by species of Mycoplasma. These include M. synoviae, M. gallisepticum, and M. meleagridis. The disease caused by M. synoviae is called infectious synovitis, causing respiratory disease and infection of the joint fluid. M. gallisepticum is also referred to as infectious sinusitis. M. meleagridis causes an infection of the air sacs. These bacteria cannot live for long outside the bird and are easily destroyed by common disinfectants. Antibiotics are not fully effective against mycoplasmosis, and once a bird becomes infected, it will always carry the bacteria. The disease is also transmitted through the egg from mother to chick. Hatcheries usually test their breeding flock for mycoplasmosis, but before purchasing, be sure to check for the disease in chicks from any hatchery.

Fowl cholera

Affects: chickens, ducks, geese, game birds, ornamental game

This infectious disease is caused by the bacterium *Pasteurella multocida*. Birds can contract this disease, which strikes suddenly and causes numerous deaths in the flock. Factors that can cause an outbreak include overcrowded pens or ponds, poor sanitation in pens, the spread of the disease from wild birds, and cold, damp weather. While sudden death is usually the first sign of the disease, some birds will have convulsions, rapid breathing, become listless, have nasal discharge, or have vents matted with droppings. Use an antibiotic in the water to treat those birds in the flock not ill from fowl cholera. All sick birds should be removed from the flock and treated elsewhere. Dead carcasses should be burned. The best way to prevent fowl cholera from harming your birds is to practice good sanitation by regularly cleaning the pen, replacing water, and checking food sources.

Pullorum disease

Affects: turkeys, game birds, ornamental game

Salmonella pullorum is the bacterium that causes Pullorum disease. It is spread from infected hens to their young through the shells and also from bird to bird through droppings. There is a high death rate among young birds affected with Pullorum disease; older birds can harbor a chronic infection. Signs of Pullorum disease include lethargy, huddling, ruffled feathers, diarrhea that pastes the vent, no appetite, and weakness. It is not too common anymore, due to nationwide efforts to eradicate the disease. If you find the disease in a bird, it should be reported to federal authorities. Treatment is unsatisfactory, and infected premises are difficult to free of the microorganism. You should purchase birds from Pullorum-free flocks and hatcheries. The hatchery should be able to produce records indicating that they test for and are free from this disease.

Paratyphoid

Affects: game birds, ornamental game

Salmonella bacteria cause paratyphoid in birds. Many different types of Salmonella cause paratyphoid. It typically causes disease in young peafowl from 1 to 4 weeks of age. Unfortunately, even if the bird survives the initial infection, they typically become chronic carriers of Salmonella, which causes diarrhea, dehydration, poor growth, and low fertility. This is another disease spread from the mother to the chick from soiled or contaminated eggs. Other sources of Salmonella include rodents and contaminated equipment or footwear. Antibiotics in the water or feed can help reduce the incidence of Salmonella.

Colibacillosis

Affects: chickens

This is an infection with *Escherichia coli (E. coli)*. The signs of colibacillosis vary. Some birds can suddenly die with no prior signs of disease, while other chickens have a chronic infection leading to a chicken with bad health. The bacteria can infect the blood, intestinal tract, or lungs, which can cause diarrhea, labored breathing, and coughing. The best way to prevent colibacillosis is through sanitation of feeders and waterers on at least a weekly basis while also keeping the pen and bedding clean. Providing good ventilation in the pen or coop will also help minimize the impact of this disease. Because there are many strains and types of E. coli, antibiotic treatment does not ensure successful treatment.

Infectious coryza

Affects: chickens

Commonly known as a cold, this occurs in adolescent and adult chickens. A chicken with coryza will have swelling of the face and nasal and eye discharge. Egg production will drop as well. Antibiotics in the water or feed can help control coryza. Prevention is aimed at keeping the organism out of the flock through making sure only healthy chickens are introduced into the flock. Purchase your chicks from a single source, and do not mix young chickens with older chickens. In fact, practicing an "all-in, all-out" management style will prevent many infectious diseases from occurring. This means the only chickens on your farm will be ones that were purchased from the same source at the same time and new additions will only occur when all the old chickens are gone.

Airsacculitis

Affects: turkeys

This is a disease that affects the turkey's air sacs. The respiratory system of a bird is drastically different from mammals. The lungs are attached to the inside of the rib cage, so there is no diaphragm to inflate and expand the lungs; instead birds have a system of thin-walled pouches, or air sacs, that connect to the lungs by openings called ostia. The muscles that attach to the breastbone cause this bone to move during breathing. This changes the pressure in the air sacs to cause the lungs to draw in and expel air. Sometimes during the process of breathing, foreign material or microorganisms can become lodged in the air sacs. These substances cause inflammation and infection of the air sacs. This can cause poor growth rate in a flock or even death losses.

To prevent airsacculitis, carefully monitor ventilation in turkey houses. You do not want ammonia fumes or dust particles to build up in the turkey barn. Opening windows or using exhaust fans will help to minimize dust particles that enter the turkey houses. Fans also release ammonia fumes produced by the decomposition of manure. Treating infected air sacs is not difficult; just add antibiotics to the turkeys' water.

Blue comb

Affects: turkeys

This disease is caused when turkeys ingest contaminated feed or water. It affects the digestive tract, leading to loss of appetite, diarrhea, decreased body temperature, dehydration, and death. Poults are the most severely affected and there is no treatment. Prevention revolves around strict sanitation of pens and equipment and keeping the young turkeys dry and warm.

Erysipelas

Affects: turkeys, game birds

Erysipelas is caused by the bacterium, *Erysipelothrix insidiosa*. The bacteria gains entry into the turkey's body through wounds, and it is found in the soil in many turkey farms. Turkeys affected with erysipelas will be lethargic, have a bluish discoloration to the head, sulfur-colored droppings, swelling of the snood, and nasal discharge. Erysipelas can be treated with antibiotics and there is a vaccination available. The disease can also be passed on to humans, swine, and sheep. Use rubber gloves when treating sick birds.

Botulism

Affects: ducks

Ducks can be affected by botulism, otherwise known as limber neck. The disease is caused by the bacterium, *Clostridium botulinum*, which grows in the mud and vegetation in warm, stagnant water. The duck ingests the bacteria, and it releases a toxin. The bird may be found dead, or it may be paralyzed and conscious. Treatment is possible during the first 24 hours by force-feeding the duck water and feed. The bird should be placed in a shaded, dry nest away from predators while the toxin wears off. The rest of the flock should be moved to a clean pen; otherwise, an outbreak can occur. To prevent botulism, replace water and feed regularly, especially if the water is warm or the feed is wet. Remove rotted scraps of food from their pen.

Salmonellosis

Affects: chickens, turkeys, ducks, geese, game birds, ornamental game

This disease is caused by the bacterium, *Salmonella*, an organism that can affect a wide variety of animals, including humans. It can quickly become a flock-wide problem due to its tendency to spread quickly. Signs of Salmonellosis include lethargy, diarrhea, swollen joints, and lameness. Identification of the disease is only made through laboratory testing of feces from infected birds and examination of carcasses of dead or dying birds. A bird that survives Salmonella will remain infected for life, and they should be separated from the rest of the flock to prevent spread of the disease. This disease can be spread to susceptible humans; therefore, practice cleanliness when working with your flock.

Parasitic diseases

Parasitic infections can plague your birds, especially if they have access to dirt. Check your birds daily to assess their overall health. Things to observe include feather loss, weight loss, unusual appearance, sitting huddled away from the flock, and decreased egg production. Infected birds can be more prone to developing other diseases that can quickly lead to death.

Most birds raised with access to dirt flooring will harbor a few internal parasites. These generally will not cause symptoms or problems and are not easily transmitted to humans.

Birds get parasites by eating the parasite eggs that can be found on food, in the dirt, or in water. Insects, earthworms, or snails — all tasty treats for foraging birds — also carry the parasites or their eggs in their bodies. To control parasites, there are some specific things you can do:

- Do not overcrowd your shed or outside pens.
- Try to keep wild birds away from your domestic flock.
- Use insecticides, if necessary, to control insects in sheds.
- Change bedding frequently and keep it dry.
- Remove droppings to keep birds from pecking at them.
- Keep your birds on a quality feed, formulated with plenty of vitamins.
- If you suspect internal parasites, have your local veterinarian identify the species so you can properly medicate the birds.

Blackhead

Affects: turkeys, game birds, ornamental game

Blackhead, or histomoniasis, is a disease that affects young turkeys and peafowl. Microscopic protozoa, *Histomonas meleagridis,* cause damage to the liver and intestine of infected birds. A bird is infected in this way: The harmful protozoa reside in cecal worm eggs; earthworms commonly eat these eggs; a bird then consumes the earthworms and becomes infected. Birds that are sick with histomoniasis will have dropped tails and ruffled feathers, and they will act dull and depressed. If histomoniasis is suspected, the treatment is with the medication dimetridazole. Any dead birds should be burned or buried to prevent contamination of the soil with cecal worm eggs. Equipment should be cleaned and disinfected as well. The soil in pens from an infected flock can remain contaminated for up to three years.

Coccidiosis

Affects: chickens, turkeys, ducks, geese, game birds, ornamental game

This disease is caused by parasites that cause decreased growth and death in birds. After ingesting the organism in feed or water contaminated with feces, it grows in the intestine and causes damage to tissues. This damage leads to decreased absorption of nutrients, decreased feed intake, blood loss, and an increased susceptibility to other infections. The primary symptoms are an outbreak of bloody diarrhea along with lethargic birds that huddle together with ruffled feathers. These outbreaks are usually related to an increased number of birds in a small space: The higher the stocking density, the greater the number of coccidia in a smaller area. With less space per bird, the chances of infection increase because there will be a greater concentration of coccidia. Levels of coccidia in the digestive tract will not cause much damage, but higher numbers will cause serious problems. It is usually a more severe problem in young birds. Treatment includes the addition of coccidiostats – medications that kill coccidia – in the feed or water. Prevention relies on manure removal, moving birds to fresh ground, and decreasing stocking density in pens.

Trichomoniasis

Affects: game birds, ornamental game

Trichomoniasis is a disease caused by Trichomonas gallinae or Trichomonas gallinarum. T. gallinae affects the upper digestive tract. Signs include drooling, difficulty eating, depression, and a foul odor from the mouth. T. gallinarum affects the lower digestive tract. This causes the affected bird to stop eating, lose weight, and have diarrhea. Infected birds need to be given medicine to kill the bacteria. To prevent infection, keep pens clean, replace food and water regularly, and do not let your birds come in contact with wild birds.

Internal parasites

Birds are quite frequently exposed to internal parasites through ingesting these organisms in the feed, water, or soil. Some common internal parasites include ascarids (round worms), cecal worms, capillaria worms, and tapeworms. Signs of a bird affected with internal parasites vary but can include pasting of the vent, diarrhea, poor or stunted growth, or lack of appetite. The only sign may be slightly reduced weight at marketing time. Worms can be treated through use of medications readily purchased at farm stores and veterinary offices.

Gapeworms

Affects: turkeys, game birds, ornamental game

Gapeworms gain entry into the bird's body when the bird eats earthworms, which harbor the gapeworm eggs. The worms become infected with the eggs when they eat the eggs passed in the dropping of infected birds. The eggs develop into worms in the bird's lungs. The growing worms block the windpipe and can cause the affected bird to suffocate. Signs include outstretched necks and coughing. The birds will not eat and will quickly become weak. Birds can be treated with anthelmintics, a deworming medicine, in the food or water. To prevent this from occurring, keep the food and water off the ground by keeping feeders on a low platform. If you have a problem, the soil in the pens can be tilled after the growing season to help destroy eggs.

External parasites

Fowl lice and mites

Affects: chickens, turkeys, game birds, ornamental game

There are numerous species of lice and mites that can affect all poultry. While lice do not directly kill poultry, they can lead to discomfort and loss of productivity. They are usually straw-colored and can be found on the skin or on feathers. Lice egg clusters are white and can be found attached at the base of the feathers. Lice spend their entire life cycle on the bird feeding on skin, feather debris, and scales. Lice are spread when an infested bird contacts other birds.

Mites can be of variable appearance, but they are generally small, crawling insects on the bird. Like the lice, they spend their life cycle on the birds and are spread from bird to bird during contact. Chiggers are a type of mite that spends the life cycle in the environment. The immature stage feeds on the lower portions of the body, causing scabby sores.

Treatment of these external parasites is through the use of an insecticide formulated for poultry and can be found at most farm stores or veterinary offices. The premises should also be treated to kill those insects hiding in cracks and crevices. You will want to treat the birds and premises at least twice, ten days apart, in order to kill lice and mites emerging from eggs.

Nutritional deficiencies

Nutritional deficiencies can be a problem in poultry, especially while they are still growing.

Rickets

Affects: game birds, ornamental game

This is caused by a deficiency of vitamin D, calcium, or phosphorus. The bones will not grow well or will become deformed in young birds. In older female birds, a similar deficiency will cause osteoporosis. This causes their bones to become fragile and prone to breaks. A lack of exercise also contributes to osteoporosis. A properly fortified feed along with exercise and free-range access to plants can help prevent rickets and osteoporosis. Having crushed oyster shells readily available will supply your birds with extra calcium.

Vitamin A deficiency

Affects: game birds, ornamental game

This results in poor weight gain, poor feather formation (which can lead to skin problems or frostbite in cold weather), and death. Adult birds deficient in vitamin A will have nasal and eye discharges and decreased egg production. Most feeds will have plenty of vitamin A, provided the feed is fresh and has not been sitting on the shelf for a long period of time. Fresh vegetation, such as grass or alfalfa, will also provide your birds with plenty of vitamin A.

Angel wing

Affects: geese

Angel wing is also known as crooked or slipped wing. The last joint of the wing becomes twisted, causing the wing feathers to point out rather than lay flat against the body. Males develop the condition more often than females. Angel wing is caused by an improper diet while the goose is young. Too much protein or too many calories combined with low vitamin D, vitamin E, and manganese levels prevent the last joint on the wing from developing properly in relation to the rest of the wing bones. There may also be a genetic predisposition to developing angel wing, and those adults affected with this disease should not be kept for breeding purposes.

The condition can affect both wings or just one wing, though if only one wing is affected, it will likely be the left wing. Feeding your geese a proper diet during the growing stage can prevent the disease. If the disease is caught early in a young, growing goose, the wing can be wrapped in wide self-adhesive wrap, called vet wrap, and the wing secured to the body. A diet with low protein and calories that is also supplemented with vitamin D, vitamin E, and manganese should be fed to help the joint develop properly in the affected goose.

Some other nutritional disorders

Affects: ornamental game

These deficiencies seen in birds include curled toe paralysis (riboflavin deficiency), crazy chick disease (vitamin E deficiency), and gizzard myopathy (selenium deficiency). As long as your feed is fresh — less than six weeks from mill date — you should not experience a problem with rations purchased by reputable feed dealers. If your birds are showing signs of a nutritional deficiency, add a powdered vitamin mix to their water or feed.

Behavioral problems

Cannibalism

Affects: chickens, game birds

While not a disease, one big cause of death among game birds raised in captivity is cannibalism. Pheasants are more prone to cannibalism, but all species and ages are capable of this vicious act. It may start as simple feather-picking and can quickly explode into a full-blown attack to the death. There are many causes of cannibalism. These include:

- Overcrowding
- Territorial aggression
- External parasites
- Nutritional imbalances
- Injuries due to poorly maintained pens or equipment
- Poor sanitation and ventilation
- Too high of a brooding temperature

To help minimize the chance of cannibalism from occurring in your game bird flock, there are a few management steps you can take, although even the best-managed flocks may still have problems. First, provide the birds with adequate floor space, shelter, and eating and drinking space. Vegetation cover will allow birds to escape and hide from aggressive birds. Maintaining the pen and equipment will cut down on chance of injuries. Remove injured, sick, or weak birds from the flock as soon as they are observed. Work quietly among your birds to avoid frightening or startling them.

Loss prevention

Discovering one of your birds is dead can be heartbreaking, but it also ruins your investment. While an occasional death will occur due to circumstances beyond your control, there are a number of steps you can take to minimize death losses and maximize bird health in your flock.

The first step is to purchase healthy stock by purchasing chicks from reputable hatcheries that test their breeding flock for diseases. Clean and disinfect your pens and equipment prior to bringing your chicks home and frequently during their stay at your farm. The steps to cleaning and disinfecting are:

1. Remove any organic material from the pen or equipment. Organic material means shavings, manure, feed, and water or any other material adhering to the surfaces of the pen or equipment.

2. Use a running stream of water, preferably under pressure to rinse the pen and equipment. In the pen, begin rinsing from the top of the pen walls to the bottom to drive any material down the walls and onto the floor. Pay special attention to corners, cracks, or crevices where any loose ma-

terial can lodge. Remove any built-up material from the pen after rinsing.

3. Use household dishwashing liquid (1/4 cup per 5 gallons of water) and a stiff brush to scrub the equipment and pens.

4. Rinse again.

5. Use a commercial disinfecting solution such as Nolvasan or Lysol (mix according to manufacturer's direction) or 1/4 cup bleach in 5 gallons of water. Apply to all surfaces of the pen or equipment and let sit for 20 minutes.

6. Perform a final rinse and allow pen or equipment to air dry. If possible, place in direct sunlight as the sunlight does have some disinfecting properties as well. Make sure every item is thoroughly dry prior to refilling or placing new bedding in the pen.

Each time you disinfect the pen area will be the ideal time to check the pen for any maintenance needs. Repair any loose or damaged boards or fencing. Check for any possible points of entry a predator may find. Remember snakes, rats, and mice can wiggle through the smallest of cracks. Use fine mesh, concrete, small-gauge welded wire, or sturdy boards to seal cracks.

If your birds are different ages, take care of the youngest group first. The older ones will most likely be harboring a few viruses, parasites, or bacteria that they have developed immunity to. Birds under 8 to 12 weeks do not have a fully developed immune system.

If you plan to have your flock on pasture, make sure there is enough ground for the birds. Overstocking pens and pastures is an invitation for disease-carrying organisms in the feces to grow and multiply. Rotating pastures through use of small paddocks will be beneficial in two ways: Fecal material will get a chance to dry, as sunlight can neutralize many disease organisms. It will also give the pasture plants a chance to grow back.

Feed your birds the best food you can afford. Check feed labels to make sure the minimal levels of nutrients are in the feed for each stage of bird growth. Store food in a waterproof and predator-proof room or container. For small flocks, a large plastic garbage can with a tight-fitting lid makes a great place to store feed. Keep wild birds, mice, and rats out of the bird-feed and bird-house. These critters tend to defecate in the feed and spread many harmful diseases.

You should check on your birds daily and record their feed and water intake. If either drops off suddenly, investigate if there is a problem with delivery or if the birds are just not eating and drinking. If they are not consuming water or feed, they may be sick. When you suspect a health problem, immediately contact a veterinarian for assistance in diagnosing the illness.

Remove any sick or dead birds from the flock immediately upon discovery. Sick birds should be isolated from the rest of the flock until they become healthy. Any dead birds should be burned, composted, or buried. Check with the local environmental office to verify which type of disposal method is allowed on your property.

Conclusion

MAKING YOUR DECISIONS EASIER

After reading, you now have learned enough about poultry breeds to decide which type of bird to raise. The breed descriptions should have helped you decide which breed to purchase based on why you are raising poultry. You should understand the basics of raising, breeding, incubating, and butchering for each type of bird. Remember, you can always refer back to the text when a question about your flock arises. *Appendix A also provides a list of resources to add to your knowledge of specific birds.*

Raising poultry is a rewarding endeavor that can teach you — and your family — new things. Do not forget to use every resource at your disposal; from this book to your local feed store, any advice can be a useful asset. Now you are ready to begin your adventure into the world of poultry.

Appendix

INFORMATION AND RESOURCES

Heritage Breed Conservation

The American Livestock Breeds Conservancy is a nonprofit organization that is working to protect endangered poultry and livestock breeds from extinction. They hope to conserve historic breeds of livestock and poultry and to maintain genetic diversity.

Here is a listing courtesy of the American Livestock Conservancy of poultry species that are in critical need of conservation or that are threatened with extinction.

Chickens

Critical	Threatened
Buckeye	Andalusian, Buttercup, Cubalaya
Campine, Chantecler, Crevecoeur	Delaware, Dorking, Faverolle, Java
Holland, Modern Game	Lakenvelder, Langshan, Malay

Critical	Threatened
Nankin, Redcap, Russian Orloff, Spanish, Sultan, Sumatra, Yokohama	Phoenix

Ducks

Critical	Threatened
Ancona, Aylesbury, Magpie	Buff or Orpington
Saxony, Silver Appleyard	Cayuga
Welsh Harlequin	

Geese

Critical	Threatened
American Buff, Cotton Patch	Sebastopol
Pilgrim, Pomeranian	
Roman, Shetland	

Turkeys

Critical	Threatened
Beltsville Small White	Narragansett
Chocolate, Jersey Buff	White Holland
Lavender/Lilac, Midget White	

To learn more, go to **http://albc-usa.org**.

Web Resources

General Farm

United States Department of Agriculture: **www.usda.gov**

The Sustainable Agriculture Research and Education Organization: **www.sare.org**

The National Sustainable Agriculture Information Service: **www.attra.ncat.org**

The National Association of State Departments of Agriculture: **www.nasda.org**

United States Department of Agriculture Food Safety and Inspection Service: **www.fsis.usda.gov**

Buildings

Free Chicken Coop Plans: **www.freechickencoopplans.com**

Feather Site: **www.feathersite.com**

PVCPlans.com: **www.pvcplans.com**

Omlet: **www.omlet.us**

Green Roof Chicken Coop: **www.greenroofchickencoop.com**

Build Eazy: **www.buildeazy.com**

Egganic Industries: **www.henspa.com**

Amish Goods: **www.myamishgoods.com**

Chicken Coop Source: **www.chickencoopsource.com**

Supplies

eNasco: **www.enasco.com/farmandranch**

Tractor Supply Company: **www.tractorsupply.com**

Horizon Structures: **www.horizonstructures.com**

Farm Tek: **www.farmtek.com**

Premier Fencing: **www.premier1supplies.com**

General Poultry

The American Livestock Breed Conservancy: **www.albc-usa.org**

University of Minnesota Poultry U: **www.ansci.umn.edu/poultry/index.html**

Chickens

National Chicken Council: **www.nationalchickencouncil.com**

My Pet Chicken: **www.mypetchicken.com**

Guinea fowl

Guinea Fowl Breeders Association: **www.gfba.org**

Frit's Farm: **www.guineafowl.com/fritsfarm/guineas**

Peacocks

Amy's Peacock Paradise: **www.amyspeacockparadise.com**

United Peafowl Association: **www.peafowl.org**

Game Birds

Pheasants Forever: **www.pheasantsforever.org**

North American Game Bird Association: **www.mynaga.org**

Feed Supply Companies

Purina Mills: **www.purina-mills.com**

Nutrena Animal Feeds: **www.nutrenaworld.com/nutrena**

Hatcheries

Porter's Rare Heritage Turkeys: **www.porterturkeys.com**

Cackle Hatchery: **www.cacklehatchery.com**

Dunlap Hatchery: **www.dunlaphatchery.net**

Holderread Waterfowl Farm & Preservation Center: **www.holderreadfarm.com**

Murray McMurray Hatchery: **www.mcmurrayhatchery.com**

Sand Hill Preservation Center: **www.sandhillpreservation.com**

Stromberg's: **www.strombergchickens.com**

Glossary

General Terms

Breed: A group with common ancestors and characteristics that distinguish the group. Often, the breeds need to be maintained through selective breeding.

Broody: Term used for hens sitting on a clutch of eggs in a nest. A broody hen will stop laying eggs to incubate the eggs and will seldom leave the nest until the eggs hatch.

Class: A smaller category of a type of birds. Often the name will relate to the origin of a bird.

Down: Young birds are born with this fluffy hair that does not protect them like feathers. The feathers grow in during the first few weeks of life.

Husbandry: The agricultural practice of breeding and raising livestock.

Hybrid: The offspring of genetically dissimilar parents.

Hygrometer: An instrument used to measure humidity. With birds a hygrometer is used to measure the humidity inside an egg incubator.

Farm: Any operation that sells at least $1,000 of

agricultural commodities or that would have sold that amount of produce under normal circumstances.

Feed Mill: A place where feed for animals is produced for commercial use.

Flighty: The tendency of a particular breed of bird to be excitable and nervous.

Free-range: When birds are allowed to graze rather than being raised in confinement.

Grain elevator: A place where farmers store and sell their grain.

Hatchery: A place where eggs are hatched.

Molt: An annual process in which a bird loses its feathers and replaces them with new feathers.

Nesting Box: A box in a pen where females can lay their eggs.

Pen: a farm building for housing poultry, also called a coop. They can be portable as well.

Pin-feathers: Also called blood feathers, these feathers contain a blood vein and if pulled will cause a bird to bleed.

Variety: In birds, this will be a unique characteristic that occurs in a certain branch of a specific breed. An example would be different feather coloring or a plume of feathers on top of the head.

Chicken Terms to Know

Bantam: A small mature chicken. True bantams are a specific breed while other bantams are smaller types of other breeds of chickens.

Broiler: Male chicken specifically bred to grow rapidly in a short amount of time. It is a young, tender chicken about 2 months old, which weighs between 2.5 to 4.5 pounds when eviscerated.

Candling: The process of shining a bright light through an egg to determine if there is an embryo developing inside the egg.

Capon: Male chickens about 4 to 8 months old that have had their reproductive organs removed. They weigh about 4 to 7 pounds.

Chick: A young chicken.

Clutch: Term used for a setting of eggs.

Cockerel: A young, immature male chicken.

Comb: The fleshy growth atop a chicken's head. Usually larger in the males.

Crop: An enlargement of the chicken's esophagus. Food is held in this pouch prior to passing into the gizzard.

Down stage: Refers to the first few weeks of life when the chick is covered in down.

Egg Bound: Occurs when the female's reproductive tract becomes blocked and the egg cannot pass from the body.

Evisceration: To remove the internal organs of a chicken being butchered.

Hen: A mature female chicken.

Pullet: A young hen.

Rooster/Cock: A mature male chicken more than 1 year of age.

Wattles: The fleshy tissue under the throat of some chicken breeds.

Turkey Terms to Know

Beard: A tuft of coarse hair that grows from the center of the breast.

Caruncle: A brightly colored throat growth.

Hen: A female turkey.

Poult: A young turkey.

Snood: A flap of skin hanging over the beak.

Spur: A hard growth on the back of the lower legs.

Tom: A male turkey.

Wattle: A flap of skin directly under the chin.

Ducks and Geese Terms to Know

Drake: A male duck. Mature drakes (except for the Muscovy breed) can be differentiated from the duck by curled tail feathers.

Duck: A female duck and the general term for the species.

Duckling: A young duck.

Gander: A male goose.

Goose: A female goose and general term for the species.

Gosling: A young goose.

Game Birds Terms to Know

Chick: A young bird.

Cock: A male bird.

Covey: A small flock of birds, especially partridge or quail.

Hen: A female bird.

Guinea Terms to Know

Chick: A young peafowl.

Cock: A male bird.

Crest: A plume of feathers on the top of the head.

Hen: A female bird.

Helmet: A distinguishing feature on the head of guineas.

Keets: A young bird.

Peacock: A male peafowl.

Peahen: A female peafowl.

Index

Author Biography

Dr. Melissa Nelson is a graduate of the University Of Minnesota – College Of Veterinary Medicine. She currently resides on a beef cattle farm near Ortonville, Minn. She is also the author of *The Complete Guide to Small-Scale Farming*, also published by Atlantic Publishing Group (**www.atlantic-pub.com**).